W0079583

# SpringerBriefs in Animal Sciences

For further volumes:
http://www.springer.com/series/10153

Muhammad Munir · Siamak Zohari
Mikael Berg

# Molecular Biology and Pathogenesis of Peste des Petits Ruminants Virus

 Springer

Muhammad Munir
Department of Biomedical Sciences
  and Veterinary Public Health
Division of Virology
Swedish University of Agricultural
  Sciences (SLU)
Uppsala
Sweden

Mikael Berg
Department of Biomedical Sciences
  and Veterinary Public Health
Division of Virology
Swedish University of Agricultural
  Sciences (SLU)
Uppsala
Sweden

Siamak Zohari
Joint Research and Development
  Unit for Virology of National Veterinary
  Institute (SVA) and Swedish University
  of Agricultural Sciences (SLU)
Uppsala
Sweden

ISSN 2211-7504          ISSN 2211-7512  (electronic)
ISBN 978-3-642-31450-6    ISBN 978-3-642-31451-3  (eBook)
DOI 10.1007/978-3-642-31451-3
Springer Heidelberg New York Dordrecht London

Library of Congress Control Number: 2012942708

© The Author(s) 2013
This work is subject to copyright. All rights are reserved by the Publisher, whether the whole or part of the material is concerned, specifically the rights of translation, reprinting, reuse of illustrations, recitation, broadcasting, reproduction on microfilms or in any other physical way, and transmission or information storage and retrieval, electronic adaptation, computer software, or by similar or dissimilar methodology now known or hereafter developed. Exempted from this legal reservation are brief excerpts in connection with reviews or scholarly analysis or material supplied specifically for the purpose of being entered and executed on a computer system, for exclusive use by the purchaser of the work. Duplication of this publication or parts thereof is permitted only under the provisions of the Copyright Law of the Publisher's location, in its current version, and permission for use must always be obtained from Springer. Permissions for use may be obtained through RightsLink at the Copyright Clearance Center. Violations are liable to prosecution under the respective Copyright Law.
The use of general descriptive names, registered names, trademarks, service marks, etc. in this publication does not imply, even in the absence of a specific statement, that such names are exempt from the relevant protective laws and regulations and therefore free for general use.
While the advice and information in this book are believed to be true and accurate at the date of publication, neither the authors nor the editors nor the publisher can accept any legal responsibility for any errors or omissions that may be made. The publisher makes no warranty, express or implied, with respect to the material contained herein.

Printed on acid-free paper

Springer is part of Springer Science+Business Media (www.springer.com)

*My first book to my first kid, Haniya Munir*
Muhammad Munir

# Foreword

I am delighted to have been asked to write the foreword for this book. The authors Muhammad Munir, Siamak Zohari, and Mikael Berg are well-qualified veterinary virologists with extensive experience on the Peste des Petite Ruminants Virus (PPRV). They have done an excellent job addressing the aspects of the PPRV research findings. I am also a veterinary virologist with more than 25 years of research experience on paramyxoviruses, a family of viruses that include PPRV.

Peste de Petits Ruminants (PPR) is a highly contagious and economically important viral disease of domestic and wild small ruminants. The clinical signs of PPR are similar to rinderpest in large ruminants, the most devastating animal viral disease known, which is now eradicated globally. PPR was first described in Côte d'Ivôire in West Africa in 1942. The disease is currently circulating in African and Asian countries and appears to be spreading into other parts of the world. It is not known whether the emergence of PPR is due to the eradication of rinderpest or due to the availability of improved diagnostic tests. There is a serious concern that the disease can spread to unaffected regions of the world and will take the place of rinderpest as one of the most widespread, destructive, and costly diseases among small ruminants.

PPRV is a member of the genus *Morbillivirus* in the family *Paramyxoviridae*. It was first thought to be a strain of the rinderpest virus but was later identified as a closely related yet distinct virus. Although significant work has been done on the molecular biology of PPRV, a reverse genetics system for the virus is still not available. Development of an efficient reverse genetics system would greatly increase our knowledge of this pathogen.

This book provides a timely and comprehensive review of the current knowledge of PPRV. To my knowledge, this is the first book to cover all aspects of the virus. The authors have done an outstanding job in compiling the latest information on PPRV. The book is organized into seven chapters covering genome organization, virus replication and determinants of virulence, pathophysiology and clinical disease, immunology and immunopathogenesis, epidemiology, diagnostic assays and vaccines, and global eradication. Each chapter is well written, clear, and informs the reader about our current knowledge on the topic.

The information in the book is well-balanced between the molecular biology and pathogenesis of PPRV. One of the most impressive aspects of the book is its broad coverage of the challenges involved in the eradication of PPR. The authors make clear that enormous progress has been made in our understanding of PPRV. However, they also concede that a great deal of basic research remains to be done so that we can understand the pathogenesis and host range of PPRV, which will be important to the success in eradicating the disease on a global level.

This publication is an invaluable reference source of timely information for virologists, microbiologists, immunologists, veterinarians, and scientists working on PPRV. It also contains scientific material appropriate for graduate and undergraduate students.

Siba K. Samal
Professor of Virology
University of Maryland
College Park MD
USA

# Contents

# Chapter 1
# Genome Organization of Peste des Petits Ruminants Virus

**Abstract** Peste des Petits Ruminants (PPR) virions are enveloped, pleomorphic particles containing a genome of single stranded RNA that is enclosed in a ribonucleoprotein core. The PPRV genome is 15,948 nucleotides (nts) long, which is the longest of all the morbillivirus members except for a recently described feline morbillivirus, which is revealed to be 16,050 bases long due to unusually long 5′ trailer sequence. The genome of PPRV encodes for eight genes in the order 3′-N–P/C/V-M-F-HN-L-5′. The mean diameter of PPR virions (400–500 nm) is slightly larger than rinderpest virus (RPV) (300 nm). As a typical feature for all members of the genus morbillivirus, the PPRV genome length follows the "rule of six", but carries a certain degree of flexibility by accommodation of +1, +2 and −1 nts, which is a unique property of PPRV among morbilliviruses. In this chapter, all of the known features of the PPRV genome structure and biology are discussed. Additionally, all of the structural and nonstructural proteins are described comprehensively.

**Keywords** Genome organization · Genome structure · Viral proteins · Virus morphology · Genome comparison

## 1.1 Introduction

Peste des Petits Ruminants virus (PPRV) is a paramyxovirus classified in the genus morbillivirus along with (rinderpest virus) RPV, measles virus (MV), canine distemper virus (CDV), phocine distemper virus (PDV), and dolphin morbillivirus (DMV). PPRV shows great structural, biological, genetic, and molecular homology to that of other members in the genus. Within the genus morbillivirus, it is MV, a human pathogen, which is extensively studied, followed by canine distemper and

M. Munir et al., *Molecular Biology and Pathogenesis of Peste des Petits Ruminants Virus*, SpringerBriefs in Animal Sciences,
DOI: 10.1007/978-3-642-31451-3_1, © The Author(s) 2013

rinderpest. However, a current surge in the molecular investigation of PPRV has made substantial contributions to the molecular understanding of PPRV, although still at the stage of infancy. Therefore, most of our current knowledge regarding PPRV is poor compared to the other members. This chapter will attempt to provide a truly comprehensive account of all of the known aspects of PPRV genome structure and organization.

## 1.2 Morphology and Genome Structure

### 1.2.1 Virion and Genomic Properties

PPR virions are enveloped, pleomorphic particles containing single stranded RNA as the genome, enclosed in a ribonucleoprotein (RNP), and core (Fig. 1.1a, b). PPRV has a genome length of 15,948 nucleotides (nts), which was considered the longest of all of the morbillivirus members (Bailey et al. 2005). However, recently a novel morbillivirus, feline morbillivirus, has been characterized from domestic cat (*Felis catus*). Complete genome sequencing of feline morbillivirus revealed that the genome is 16,050 bases, the largest among morbilliviruses identified so far, because of unusually long 5′ trailer sequences of 400 bases (Woo et al. 2012). This is a single report and requires further characterization before its full valida-tion. Due to the variable length of the intergenic region between the M and F genes (without having an effect on the protein lengths), the length of the genome varies in morbilliviruses. So far, no obvious role for this variable and high GC content intergenic region has been observed in the replication of the morbilliviruses. The mean diameter of PPR virions (400–500 nm) is also larger than RPV (300 nm) (Gibbs et al. 1979). Negative staining in the electron microscope revealed that the thickness of the PPRV envelope varies from 8–15 nm while the length of the spikes ranges from 8.5–14.5 nm. The ribonucleoprotein strands appear as a herring bone carrying the thickness of ∼ 14–23 nm (Durojaiye et al. 1985).

As a typical feature for all the members of the genus morbillivirus, the PPRV genome length also follows the "rule of six", meaning that their length is poly-hexameric, which is required for the efficient replication of the viral genome (Kolakofsky et al. 1998). Contrary to this strongly accepted belief, a recent study revealed that PPRV obey the rule of six but carry a degree of flexibility (Bailey et al. 2007). By a still unknown mechanism, transcription and replication in PPRV can accommodate some deviation in genome length, such as +1, +2, and −1 nts.

The PPRV genome contains six transcriptional units that encode for six con-tiguous and nonoverlapping proteins. All of the genes in PPRV are arranged in an order of 3′-N–P/C/V-M-F-HN-L-5′ and each gene is separated by an intergenic region of variable lengths (Diallo 1990) (Fig. 1.1c). The genomic RNA is sur-rounded by the nucleoprotein (N) to form the nucleocapsid, into which two other viral proteins are coupled: the large protein (L) and the phosphoprotein (P)

**(a)**                                                                   **(b)**

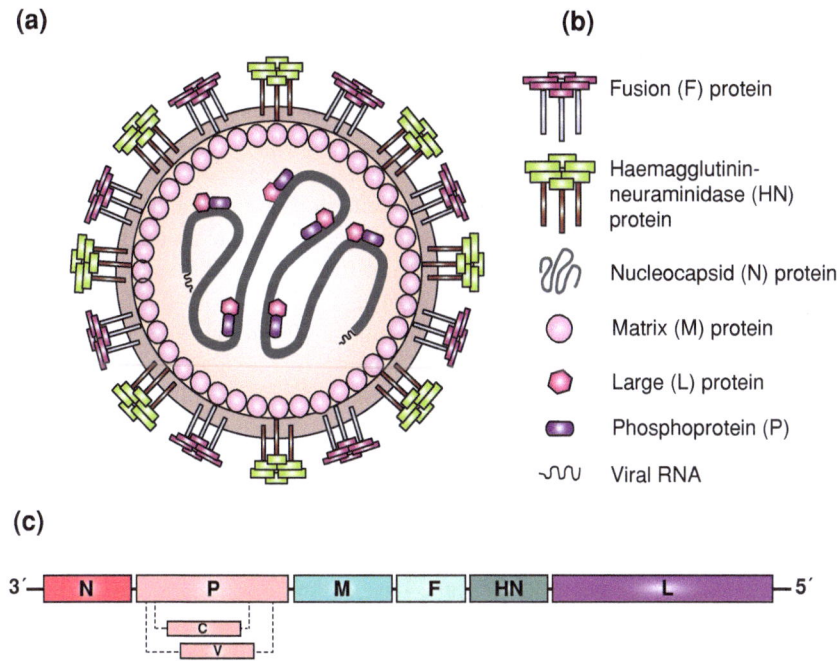

Fig. 1.1  Schematic structure of the PPR virion. **a** The PPRV genome consists of a single strand, which is surrounded by a host-derived envelope. In the virions, the P, N, and L proteins (out of total 8 proteins) constitute the nucleocapsid that encloses the viral genome, while HN and F are the spike glycoproteins that, with the association of M protein, form the viral envelope. **b** The key for the names of each protein and viral components. **c** The PPRV genome is 15,984 nucleotides long and encodes for eight proteins, in which each gene codes for a single protein except the P gene. This gene is transcribed into two nonstructural proteins (C and V) in addition to the P protein. The genome organization is shown from 3′ to 5′ end

(Fig. 1.1a, b). The P acts as a co-factor of L, which is the viral RNA dependent RNA polymerase (RdRp). There are three proteins associated with the host cell membrane derived viral envelope. The M protein acts as a link, which associates with the nucleocapsid and the two external viral proteins, the fusion (F) protein and the hemagglutinin-neuraminidase (HN) protein. Like other morbilliviruses, PPRV also encodes two nonstructural proteins, V and C, where V uses an RNA editing strategy from the P gene transcription unit (Barrett et al. 2006). The C protein is translated from an internal coding frame transcript using the second AUG start codon.

The 3′ and 5′ untranslated regions (UTRs) are crucial for the transcription and replication of paramyxoviruses, and are known as genome promoter (GP) and anti-genome promoter (AGP), respectively (Lamb and Kolakofsky 2001). In PPRV, the 3′-genome terminus, a seat for the attachment of the RdRp polymerase complex, is a stretch of 107 nts, which includes the 52-nt leader region, and 3′ UTR of the N gene, both separated by a trinucleotide GAA. This stretch of GP before the N gene's open reading frame (ORF) start codon acts as a promoter for both the synthesis of viral RNA and the production of a full-length positive sense antigenome RNA copy. A gene start and polyadenylation signal is located 52 nt downstream of the N ORF stop in PPRV, which is highly conserved among the morbilliviruses (Bailey et al. 2005).

The AGP, which is responsible for the synthesis of genome-sense RNA, is the complement of the 5' UTR after the L protein stop codon, including the trailer region that becomes the 3' end of the antigenome. A bipartite model has indicated two distinct domains required for the efficient functioning of Sendai virus (SeV), another member of the same *Paramyxoviridae* family in both the GP and AGP regions (Hoffman and Banerjee 2000; Barrett et al. 2006). The conserved 3' and 5' termini in the family reflect the similarity in their promoter activities lying in these regions. A nt stretch of 23–31 at the 3' terminus of both the GP and the AGP in PPRV is conserved and needs to be shown as an essential domain required for promoter activity. This region is believed to interact with a conserved area comprising a succession of three hexamer motifs (CNNNNN). Although the exact mechanism of how these two domains interacts and function remains unclear, a model has been proposed which predicts that the three hexamer motifs in the second promoter element lie on the same face of the helix, exactly above the first three hexamers at the 3' terminus (Lamb and Kolakofsky 2001). It is, therefore, more likely that these two regions in the GP and AGP interact directly with each other to form a functional promoter unit. A similar assembly is also presented in the promoters of the other paramyxoviruses (Murphy and Parks 1999). At the junction of the GP and N gene start, a conserved intergenic triplet sequence is also considered necessary for transcription (Mioulet et al. 2001). A recent study conducted by Bailey et al. (2007) tried to figure out the role of GP and AGP by using chimeric minigenomes of PPRV and RPV (Bailey et al. 2007). They indicated that the use of PPRV AGP decreased the ability of RPV to rescue the chimeric minigenome, which predicts the difference in closely related viruses. AGP is a very strong promoter and is responsible for a single function: the production of the full-length negative sense genome. The GP has two functions: transcription of virus mRNAs and transcription of the full-length positive sense virus genome.

## 1.2.2 PPRV Structural Proteins

### 1.2.2.1 Nucleocapsid (N) Protein

The nucleocapsid (N) protein is one of the main structural proteins, especially in nonsegmented negative stranded RNA viruses. The N protein does not induce protective immunity against the virus, but it is the most abundant and most immunogenic among PPRV proteins. This is why the N protein has been used extensively in the development of diagnostic tests. Recently, Dechamma et al. (2006) have identified the most immunogenic region [amino acid (aa) 452–472] in the PPRV N protein (Dechamma et al. 2006). Besides immunogenicity, classification of PPRV into four lineages based on the N gene better reflects the geographical origin than does variation of the external glycoproteins F and HN of PPRV (Diallo et al. 2007). The ORF for the N protein in PPRV starts at position 108 (UAC) and ends at 1685 (AUU), which upon translation produces an N protein of 58 kDa.

Based on sequence similarity of PPRV with other morbilliviruses, the N protein can be divided into four regions. Region I is the well conserved sequence (aa 1–120) carrying 75–83 % identity across the group, while region II (aa 122–145) is only 40 % identical. Region III (aa 146–398) and region IV (aa 421–525) have the most and the least conserved sequences, respectively (Diallo et al. 1994; Bailey et al. 2005). Recently, a study conducted by Choi et al. (2005) indicated that region I and II are more immunogenic than are region III and IV in the PPRV N protein (Choi et al. 2005), which corresponds to other morbilliviruses including MV (Buckland et al. 1989) and RPV (Choi et al. 2003). Furthermore, the humoral immune response occurs earlier against region I than region II. It is now well-known that the N protein is the main cross-reacting antigen among morbilliviruses. Based on monoclonal antibodies, it has been demonstrated that the N protein is somewhat different in closely related viruses (Bodjo et al. 2007), which is according to a general speculated that RPV is the archevirus from which other related viruses originate, CDV was first and PPRV was last among others (Norrby et al. 1985).

The N protein influences the virus life of paramyxoviruses cycle at multiple steps: association with the M protein facilitates virus assembly, a requirement to encapsidate the genomic RNA, and is involved with the P-L polymerase complex during replication and transcription. Recently, an in vitro study conducted by Servan de Almeida et al. showed the role of the N protein in the replication of the PPR virus by silencing the N gene (Servan de Almeida et al. 2007). Because all of the viral mRNAs are synthesized from the promoter region at the beginning of N gene, it was expected that targeting the N gene would reduce viral replication. They further revealed that the inhibition of the N protein also has a negative role on the yield of the M protein. The following year, Keita et al. (2008) further narrowed down the active site in the N gene, and presented the 5'-RRWYYDRNUG-GUUYGRG-3' motif (where R is A/G, W is A/U, Y is C/U, D is G/A/U and N is any base), silencing of which leads to inhibition of N transcript in PPRV, RPV, and MV (Keita et al. 2008). Specifically, in PPRV this translated motif (RINWFEN) is located at position 143–149 in the N protein, and the central part (NWF) is conserved among strains of each member. This study further claims that the inhibition of N transcript with subsequent inhibition of the M transcript resulted in the inhibition of PPRV replication by 10,000-fold compared to nonsilenced ones.

In morbilliviruses, the N protein can be divided into two parts called $N_{CORE}$ and $N_{TAIL}$. The $N_{CORE}$ is 420 amino acids long and is very conserved, not only among morbilliviruses but also among PPRV strains, probably reflecting its vital function in nucleocapsid assembly. In PPRV, there is a sequence stretch F-X4-Y-X4-SYAMG (X is any amino acid) at residue 324 in $N_{CORE}$ involved in the N–N self-assembly and N-RNA interaction. The $N_{TAIL}$, a 12 kDa C-terminal domain (CTD), is a relatively less conserved area in both morbilliviruses and in PPRV strains. CTD residues 488–499 are responsible for the interaction between N and other viral proteins such as P and P-L polymerase complex (Karlin et al. 2002; Kingston et al. 2004). It is believed that CTD is exposed on the surface of the protein and is easily cleaved by trypsin digestion. Despite the removal of $N_{TAIL}$, $N_{CORE}$ retains the ability to form nucleocapsid-like structures in MV (Giraudon et al. 1988). Karlin et al. (2002) observed that mutation at 228S and 229L in the MV, N protein leads to impaired self-association, and is unable to package RNA, suggesting that correct self-polymerization of the MV N protein may create a structure involved in RNA binding (Karlin et al. 2002). Sequence analysis of the PPRV nucleocapsid protein also revealed strong similarity at this region, and the presence of the same amino acids at both positions might reflect a similar function in PPRV. The morbillivirus N protein interactions with host regulatory proteins, such as heat shock protein Hsp72, interferon regulatory factor IRF3, and cell surface receptor, explain the role of the N protein in virus replication and cell tropism (Zhang et al. 2002; Laine et al. 2003). Because of the N protein sequence similarity of PPRV with other members of morbilliviruses, it is likely to have functional homology within the members of the genus.

Binding of N and P proteins to the viral leader and trailer part of the genome starts viral genome assembly, which is then followed by N–N and N-RNA interactions. This N–N self-association has only been well-studied in MV. PPRV is the second morbillivirus where this N–N interaction has been investigated recently (Bodjo et al. 2008). It has been demonstrated that two domains, one at the N-terminus (1–120) and the other in the central region (146–241), are responsible for the PPRV N–N self-assembly. They further explained that a short fragment in the N protein at aa 121–145 is essential for the stability of this nucleocapsid structure. This domain is much conserved among morbilliviruses (Diallo et al. 1994).

Although morbilliviruses replicate in the cytoplasm of the cell, the N protein forms nucleocapsid-like aggregations to make a more condense shape, which can be seen in both the cytoplasm and nucleus of mammalian transfected cells (Huber et al. 1991). Nuclear localization signal (NLS) and nuclear export signal (NES) have been identified in MV, RPV, and CDV N proteins (Sato et al. 2006), which is now further extended to PPRV. The NLS and NES presented in CDV are very similar to that of PPRV: these are TGILISIL and LLRSLTLF, and TGVLISML and LLKSLALF, in CDV and PPRV, respectively. The NLS motif is found at position 70–77 and NES was observed at 4–11 in both CDV and PPRV. Despite the fact that the PPRV strain Nigeria75/1 carries a similar motif, a preliminary study indicated that neither the naïve N nor the corresponding recombinant protein is found in the infected cell nucleus (Barrett et al. 2006). Contradictory to this, another study showed, using immunofluorescence antibodies technique, that Nigeria75/1 and Nigeria76/1 were found both in the cytoplasm and nucleus of infected cells (Chard et al. 2008). The nuclear translocation inability of Nigeria75/1 might explain the transient nature of this strain; however, further studies using specific antibodies against N proteins will clarify this localization and its possible influence on the virus replication in PPRV infection.

### 1.2.2.2 Phosphoprotein

In PPRV, nts 1807–3333 encode for P proteins that upon translation produce a protein of putative molecular weight 60 kDa, while the P protein from infected cells migrates as 79 kDa on SDS-PAGE (Diallo et al. 1987). This size variation from 72–86 kDa for most of morbilliviruses is due to its acidic nature and post-translational phosphorylation of the protein that is rich in serine and threonine (Diallo et al. 1987). There are five potential phosphorylation sites as predicted by Netphos 2.0 (Blom et al. 1999), and all are conserved among morbilliviruses. Four of the corresponding sites at aa 151, 307, 361, 470 are also conserved within PPRV, but site 348 is threonine in PPRV instead of serine. The functional importance of phosphorylation in the P protein is not completely understood. This is partially due to a lack of information on the exact phosphorylation sites, and that it has been observed that the correlation between intracellular and cell free phosphorylation does not exist (Shiell et al. 2003). Although for the P or V protein phosphorylation appears not to be a prerequisite for viral transcription, replication, and pathogenesis, information regarding the phosphorylation status in PPRV is still required. There are three serine residues at 49, 38, and 151, considered as potential phosphorylation sites, but only serine at 151 is conserved in all the morbilliviruses including PPRV (Kaushik and Shaila 2004). Although putative amino acid sequence analysis of the P protein indicates only 47 % similarity between MV and PPRV, their C-termini are more similar than their N-termini. The N protein in complex with the P protein could only be found in cytoplasm, while the nascent form can be seen in both cytoplasm and nucleus (Gombart et al. 1993).

In paramyxoviruses, the P protein performs multiple functions. In MV, interaction of the C-termini of the N and P proteins is intrinsically unstructured and unstable, and this rapid transition state facilitates the copying of the template RNA and encapsidation of the nascent RNA during replication. This N–P interaction is also required for key biological processes such as cell cycle control, transcription, and translation regulation (Johansson et al. 2003). There are two motifs in the RPV P protein that are responsible for the N–P interaction. One is at the amino terminal at position 1–59 while the other is at the carboxyl terminal at position 316–346. The P protein is the vital element of the viral L-polymerase complex, and it is assumed to be key determinant of cross-species morbillivirus pathogenicity (Yoneda et al. 2004). Oligomerization rather than phosphorylation is required for P protein activity in transcription/replication with the RdRp complex (Rahaman et al. 2004). Despite of these vital roles of P protein in the replication of morbilliviruses, its function in PPRV replication and pathogenicity remained elusive, which warrant future investigation.

### 1.2.2.3 Matrix (M) Protein

The ORF for the PPRV M protein is located at nt position 3,438–4,442, which is translated to a protein of 335 amino acids with a predicted molecular weight of 37.8 kDa. The M gene starts with a characteristic AGGA sequence at nt 3,406 and ends with AAACAAAA. This gene end is conserved in MV, RPV, and in all PPRV strains (Muthuchelvan et al. 2006). It is believed that binding of the M protein with actin filaments, a potential site for MV assembly and cellular transport, leads to apical budding of MV which is then excreted by the host to infect neighboring cells (Riedl et al. 2002). The M protein also interacts with the N protein and cytoplasmic tails of the H and F proteins (Baron et al. 1994). Moreover, it is known that PPR virus particles are released from the microvilli of intestinal epithelial cells and are therefore shed in faeces (Bundza et al. 1988). Recently, a motif FMYL has been identified in Nipah virus, a member of the same family, at position 50–53 which is believed to be required for localization of M to the cell membrane and budding process (Ciancanelli and Basler 2006). PPRV also contains an exactly similar domain at the same position, and its function might correlate to that of the Nipah virus. There is 92–99 % sequence similarity among PPRV strains; however, functional domains within M proteins remained to be investigated. A 1,080 nt long nonconserved UTR exists at the junction of the M and F genes, which includes the gene end of M and gene start of F gene, is rich in GC content. The functional relevance of this region is yet to be clarified, but a study conducted by Takeda et al. (2005) showed that this long $3'$ UTR in the MV M protein has a role in upregulation of the M protein, while the long $5'$ UTR in the F gene causes a decrease in F protein production (Takeda et al. 2005). The three ATG repeats (tctATGATGATGtca) identified in PPRV at position 991–999 are also found in the M protein of RPV and MV while the M gene of CDV, PDV, DMV lack this domain (Muthuchelvan et al. 2006). Taken together, these factors may alter cytopathogenicity of the virus in host specific manner.

### 1.2.2.4 Fusion (F) Protein

The F gene is very conserved among PPRV strains, and it encodes for a GC-rich protein 546 amino acids long with a predicted molecular weight of 59.137 kDa. This high level of sequence conservation might be the explanation for extensive cross-protection among and between different genera of morbilliviruses; for example, the vaccine against RPV can be used to immunize animals against PPRV. The sequence homology among paramyxoviruses also explains the common fusion property and hence their conserved F protein-based biological activity. The long $5'$ UTR (628 nt in PPRV), as in other morbilliviruses, is believed to be rich in secondary stem loop structures and contains potential initiation codons before the actual codon used for F protein translation. PPRV also contains a long $3'$ UTR of 136 nt which ends at AAACAAAA, followed by an intergenic trinucleotide CTT (Dhar et al. 2006). The F and H proteins are embedded in the viral lipid bi-layer envelope and protrude as spikes (Fig. 1.1a). In morbilliviruses, the F protein, with the help of the H protein, as a fusion promoter, mediates viral penetration into mammalian cells by fusing the viral and cellular membranes at the cell surface, after which the viral genome gains access to the host cell (Moll et al. 2002).

$F_0$, an inactive precursor, is one of the key molecular determinants of para-myxovirus virulence, which mainly depends on the amino acid sequence of the cleavage site and the ability of cellular proteases to cleave the F proteins. Although this cleavage is not essential to assemble the virus, it is a prerequisite for viral infectivity and pathogenesis (Watanabe et al. 1995). Under post-translational proteolytic cleavage of $F_0$, two active subunits $F_1$ and $F_2$ will be produced, which remain linked to each other by disulfide bonds. Sequence analysis of the $F_0$ protein reveals high conservation among morbilliviruses except for two variable hydro-phobic domains. The first (N-terminal) domain is responsible for bringing the proteins to the rough endoplasmic reticulum for translation while the second (C-terminal) domain is associated with anchoring the protein in the membrane. The latter is believed to remain on the cytoplasmic side of the membrane, where it can interact with the M protein and facilitate budding, because the mutation in this region leads to inhibition of virus production (Meyer and Diallo 1995). These domains are highly conserved among all the PPRV strains. At the cleave site $RRX_1X_2R$ ($X_1$ indicates any amino acid but $X_2$ must be either arginine or lysine) proposed for the morbilliviruses, PPRV carries RRTRR at position 104–108, which is recognizable by the *trans*-Golgi associated furin endopeptidase (Chard et al. 2008). In paramyxoviruses, the membrane-anchoring subunit of F1 contains four well-described conserved motifs: an N-terminus fusion peptide (FP), heptade repeat 1 (HR1), HR2, and transmembrane (TM) domain. In PPRV, the 3-D structure of the HR1-HR2 complex has revealed that heterodimer between HR2 and HR1 covers the inner core of HR1 trimer, resulting in a six-helix bundle (Fig. 1.2). Upon anchoring the FP domain in the membrane, dimerization of the HR domains leads to fusion between the host cell membrane and the viral enve-lope by bringing them close to each other (Rahaman et al. 2003). Because most of the paramyxoviruses, such as SV5, NDV, and PPRV, carrying the same structure of these hepated repeat, it is likely that they have a common fusion mechanism. In the paramyxoviruses, the F proteins contain a leucine zipper motif, which in PPRV is located at position 459–480 and is conserved among PPRV strains. This motif is responsible for facilitating the oligomerization and fusion function of the F protein through an unknown mechanism (Plemper et al. 2001).

As a typical feature of all membrane-associated proteins, the F protein under-goes potential N-linked glycosylation. In this process of post-transcriptional modification, addition of the oligosaccharide side chain is critical for the transport of the protein to the cell surface, to maintain its fusogenic ability and integrity. All members of the morbillivirus genus contain a conserved NXS/T (X indicates any amino acid) glycosylation site in the $F_2$ subunit of the mature protein (Meyer and Diallo 1995). In PPRV, the three N-linked glycosylation sites are NLS, NIT, and NCT at amino acid positions 25–27, 57–59, and 63–65, respectively. Their specific functions still need to be revealed.

**Fig. 1.2** 3-D structural model of HR1-HR2 complex of PPRV F protein. Green ribbons represent the HR2 while blue and red ribbons indicate the HR1 chains. Balls and sticks show the interaction between HR1-HR1 and HR1-HR2 in the model. Adopted from Rahaman et al. (2003), with permission

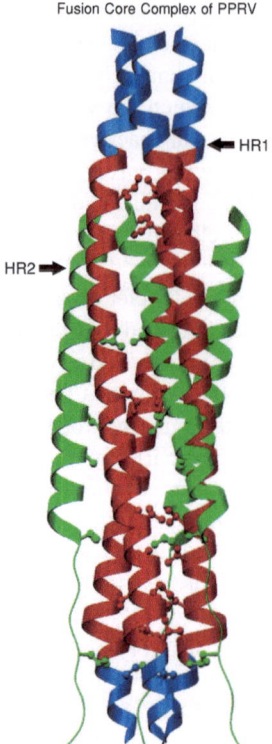

Fusion Core Complex of PPRV

HR1

HR2

### 1.2.2.5 Hemagglutinin-Neuraminidase (HN) Protein

The HN or H is one of the least conserved proteins among morbilliviruses, where two closely related ruminants viruses, RPV and PPRV, show only 50 % amino acid identity while both have 609 amino acid residues in their proteins. These variations probably reflect the specificity for cell tropism and that the host's humoral immune response is mainly directed against the HN protein. Mapping the epitopic sites on the H protein in a B cell indicated its immunodominancy, and hence it is under continuous increased immunological pressure due to the production of neutralizing antibodies (Renukaradhya et al. 2002). In PPRV strain Nigeria 75/1, the ORF ranges from 7,326–9,152 nt which encodes for a 67 kDa HN protein. In MV, the H protein mediates the virus binding to host cellular receptors, a first step in the progression of virus infection. The HN protein is a disulphide-linked homodimer in which the N-terminus, a signal peptide, appears at the cytoplasmic side of the membrane while the C-terminus protrudes on the outer side of the membrane (Vongpunsawad et al. 2004). The H protein is a major determinant of cell tropism in MV and the main cause of cross-species pathogenesis in lapinized RPV (Yoneda et al. 2002), collectively indicating that H is the vital antigenic determinant of the morbilliviruses. In an attempt to formulate a chimeric RP vaccine to avoid confused detection of PPR and RP, Das et al. (2000) have created cDNA copies of the RPV in which either HN or F or both genes were replaced with the PPRV corresponding genes (Das et al. 2000). The chimeric RPV, carrying either of the glycoproteins, could not be rescued, probably indicating the importance of type-specific interactions among these glycoproteins that leads to viral entry, assembly, and budding (Das et al. 2000). It was only possible to rescue the chimeric virus when both HN and F proteins of PPRV were replaced in the genome of RPV, and such recombinant virus grew more slowly in tissue culture than either parental virus and formed abnormally large syncytia. Despite this, the virus-induced immunity in goats to protect them from wild-type PPRV. However, there are certain observations, contradictory to this hypothesis, in which the H protein from MV and RPV are interchanged with CDV (Brown et al. 2005).

Out of the few morbilliviruses (RPV, MV, CDV, and PPRV), it is only MV in which the cell surface receptors CD46, which is used by cell cultured adapted vaccine strains, and SLAM (signal lymphocyte activating molecules, also called CD150), which is used by both vaccine and wild-type viruses, are well studied (Tatsuo et al. 2000). Despite having relatively high structural (60 % at aa level) and functional similarity with human SLAM, mouse SLAM is unable to act as a receptor for MV. It is believed that a conserved motif from amino acid 58–67 (especially isoleucine at 60, histidine at 61, and valine at 63) is critical for the difference between human and mouse SLAM receptors (Ohno et al. 2003). Similar amino acids in caprine (goat, a natural host of PPRV) SLAM at position 60 (isoleucine) and at 61 (histidine) have been identified, and might be crucial for the susceptibility of this host to PPRV infection. Another study indicated that there are seven residues that are vital for the interaction between human SLAM and MV, and six of them (Y529, D530, R533, F552, Y553, and P554) are conserved within PPRV (Vongpunsawad et al. 2004). Consistent with these studies, Pawar et al. (2008) confirmed that SLAM can act as a co-receptor for PPRV using the siRNA technique (Pawar et al. 2008). Under silenced SLAM receptor in B95a cells (a marmoset lymphoblastoid cell line), PPRV replication was observed to be reduced by 12–143 fold, while the virus titer ranged from $\log_{10}$ 1.09 to 2.28 (12–190 times) (Pawar et al. 2008). Indeed, other putative receptors for PPRV are still waiting to be identified. Asparagine (N) linked glycosylation in PPRV was found to be at N18KTH21, N172KSK175, N215VSS218, N279MSD282, and N215VSS218, as predicted using the ScanProsite program (Gattiker et al. 2002; Dhar et al. 2006).

In some of the paramyxoviruses, glycoproteins perform not only hemagglutination but also a neuraminidase function. However, among morbilliviruses only MV and PPRV have hemagglutination capabilities (Varsanyi et al. 1984; Seth and Shaila 2001). PPRV is unique for its neuraminidase activity, and therefore is the only member of the morbilliviruses that has HN protein (Seth and Shaila 2001). RPV, a very close virus to PPRV, has limited neuraminidase activity but cannot act as a hemagglutinating agent for the erythrocytes (Langedijk et al. 1997). Renukaradhya et al. (2002) using mAbs, have mapped the functional epitopes in the HN protein of PPRV (Renukaradhya et al. 2002). Two regions, one at aa 263–368 and another at aa 538–609, were identified as immunodominant epitopes.

It is now believed that the HN protein in morbilliviruses (especially PPRV) is not only responsible for viral attachment to cell surfaces and agglutination of eryth-rocytes (hemagglutination activity) but also cleaves sialic acid residues from the carbohydrate moieties of glycoproteins (neuraminidase activity), which was previously thought to be absent.

### 1.2.2.6 Large (L) Protein

Among morbillivirus proteins, the ORF for the L protein (6,949 nt) encodes for 2,183 amino acids, which is the largest protein; and, due to attenuation at each gene junction, mRNA encoding for the L protein is the least abundant (Flanagan et al. 2000). Notably, the L protein is conserved among morbilliviruses: PPRV has an identity with RPV and CDV of 70.7 and 57.0 %, respectively (Bailey et al. 2005) (**Table** 1.1). The length (2,183 aa) and the predicted molecular weight (247.3 kDa) of the PPRV L gene is comparable with other morbilliviruses (RPV, MV, and DMV), but the charge of +14.5 in PPRV is different from RPV and PDV, where it is +22.0 and +28.0, respectively. Like other –ssRNA viruses, PPRV L gene is also rich in isoleucine and leucine contents (18.4 %) (Muthuchelvan et al. 2005).

The L protein, being RdRp, is responsible for the transcription and replication of genomic RNA, including initiation, elongation, and termination. The L protein is also liable for the capping, methylation, and polyadenylation of viral mRNA. The PPRV L gene start motif (AGGAGCCAAG) is in agreement with the gen-eralized morbillivirus L gene start motif [AGG(A/G)NCCA(A/G)G]. Two func-tions of the L protein, generation of viral L gene mRNA ,and signal for the capping, start with the recognition of this motif. As with all morbilliviruses except MV, the initial codon is flanked with the Kozak sequence motif [(A/G)CCAUG] which is responsible for the start of translation in eukaryotic cells (Kozak 1986).

Studies indicate that the L protein from negative-sense nonsegmented RNA viruses can be divided into three independent domains, separated by two variable areas (at nt 607–652 and 1695–1717), where each domain carries diverse functions (Malur et al. 2002; Cartee et al. 2003). Of these three domains, the first two are positively while the third is negatively charged. The first N-terminus domain of 1–606 residues carries an RNA binding motif, KEXXRLXXKMXXKM (X indi-cates any amino acid) that is highly hydrophobic in nature, and has K rhythmically spaced with basic amino acids. In PPRV, this motif KETGRLFAKMTYKM at amino acid position 540–553 corresponds to that in other morbilliviruses. The negativity in the domain may be linked to its ability to bind with RNA. It is also interesting to observe that the part of the N and P proteins that are linked to bind with the RdRp complex are also negatively charged. In PPRV this domain also carries an invariant peptide GHP from 357–359 aa, which perhaps constitutes a turn structure and performs an important function with exposed histidine residues in all morbilliviruses (Poch et al. 1990). In the second domain (650–1694 aa), there are two motifs, one at position 771 (QGDNQ) and the other at 1,464 (GDDD),

**Table 1.1** Nucleotide (nt) and amino-acid (aa) comparison for open reading frames of each gene of PPRV (Turkey 2000, accession no. AJ849636) with other morbilliviruses

| Virus strain (Accession no.) | Rinderpest Kabete 'O' (X98291) (%) | | Measles virus 9301B (AB012948) (%) | | Canine distemper virus Onderstepoort (AF305419) (%) | | Dolphin morbillivirus CeMV (AJ608288) (%) | | Post-translational modification |
|---|---|---|---|---|---|---|---|---|---|
| Protein/level | nt | aa | nt | aa | nt | aa | nt | aa | – |
| Nucleocapsid protein | 66.2 | 72.9 | 66.8 | 73.5 | 62.5 | 68.5 | 66.2 | 72.9 | Glycosylation |
| Phosphoprotein | 62.4 | 50.5 | 60.4 | 45.1 | 56.6 | 45.3 | 61.6 | 49.1 | Phosphorylation |
| C protein | 58.8 | 41.8 | 53.2 | 40.3 | 53.1 | 35.0 | 58.5 | 37.2 | – |
| V protein | 61.9 | 45.1 | 59.0 | 41.3 | 54.2 | 40.4 | 59.0 | 43.5 | Phosphorylation |
| Matrix protein | 72.2 | 66.1 | 74.0 | 68.2 | 73.1 | 60.3 | 69.1 | 64.0 | – |
| Fusion protein | 68.0 | 73.8 | 67.0 | 71.7 | 50.2 | 54.2 | 65.9 | 73.3 | Glycosylation |
| Hemagglutinin/ Neuraminidase | 55.5 | 39.4 | 53.5 | 34.5 | 47.2 | 28.4 | 52.9 | 37.3 | Phosphorylation |
| Large protein | 68.1 | 75.6 | 68.1 | 75.1 | 64.8 | 71.1 | 67.3 | 73.7 | Glycosylation |
| Complete genome | 63.7 | – | 63.4 | – | 58.5 | – | 62.0 | – | – |

The complete genome row represents the full-length comparison among morbilliviruses. Similarity scores were calculated by using BioEdit version 7.0.9.0

which are believed to be involved in the RNA polymerase functional sites (Blumberg et al. 1988). In this pentapeptide (QGDNQ) motif, GDN tripeptide is similar to the Asp–Asp (GDD) domain of +ssRNA polymerases, and is conserved among morbilliviruses. It is believed that this GDN sequence is responsible for phosphodiester bond formation, template specificity, and cation binding (Muthuchelvan et al. 2005; Poch et al. 1990). The third domain carries not only an ATP binding site at 1,788 amino acid (a stretch of GXGXGX followed by lysine rich region, in PPR it is GEGSGS) but can also perform kinase activity, although its functional role is not clear yet (Blumberg et al. 1988). Recently, Muthuchelvan et al. (2005) claim that, like other −ssRNA viruses, the L protein of morbilliviruses can also be divided into six domains, with two regions having low similarity (residues from 607–650 and 1695–1717) (Muthuchelvan et al. 2005). This difference in the prediction of the functional domains might be due to differences in the algorithm of each domain prediction program.

In morbilliviruses, the L protein can only perform its function as an RNA polymerase when it associates with the P protein. The sequence ILYPEVHLDS-PIV at positions 9–21 can act as a binding site for P and L proteins (Horikami et al. 1994). This motif is predicted to be a coiled, surrounded by a α helix and β sheet. Cevik et al. (2003) proposed that the association of L2P4 or L2P8 is critical in the RdRp complex, and this picture would be even more complex when interaction of the V and C proteins, which are expressed from the P gene, in this complex is taken into account (Sweetman et al. 2001) This sequence for P-L interaction is conserved in paramyxoviruses: in PPRV it is totally conserved except the first

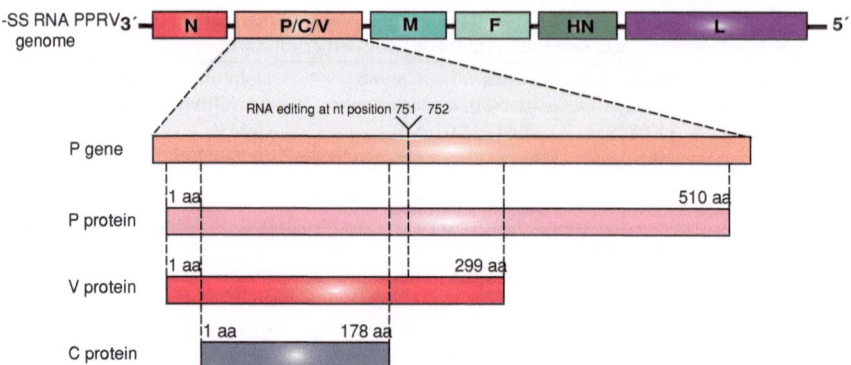

**Fig. 1.3** The P gene of PPRV, as for other paramyxoviruses, not only encodes for the usual P protein but also for two nonstructural proteins in infected cells. The mRNA for two accessory proteins, C and V, is transcribed through alternative reading frames and RNA editing, respectively

amino acid, which is valine instead (Chard et al. 2008). Despite the fact that both amino acids (V and I) are hydrophobic, and hence less likely to make any difference in interaction, their contribution in interaction between L and P proteins is not practically investigated.

## *1.2.3 PPRV Accessory Proteins*

Paramyxoviruses not only encode for the six structural proteins but also for two nonstructural proteins in infected cells. Among the morbilliviruses, all mRNAs encode for a single protein except the transcripts from the P gene. The P gene encodes for C and V gene, which are mediated through alternative open reading frame and RNA editing, respectively, in addition to collinear P gene (Fig. 1.3).

### 1.2.3.1 C Protein

The C protein is translated from the P gene and is generated from the second initiation codon (ATG at nt 82) within the P ORF (Fig. 1.3). The PPRV C protein is a short protein that consists of 117 residues with a predicted molecular weight of 20.11 kDa, which is the same length as in RPV but is three amino acids longer than CDV and PDV (Barrett et al. 2006). Among morbilliviruses, the MV C protein is the longest (186 aa) while DMV is the shortest one (160 aa).

Unlike the P protein, the C protein is not phosphorylated and can be detected in both cytoplasmic and nuclear compartments in MV infected cells (Bellini et al. 1985), while in RPV infections the C protein is only detected in the cytoplasm (Sweetman et al. 2001). Inconsistent results indicate that association of the C protein with the L protein can modulate RdRp functionality in RPV (Sweetman et al. 2001), while others found no interaction with other viral proteins (Liston et al. 1995). Its interaction with RNA can be suggested due to its strong positive charge at physiological pH. Recently, the C protein of the RP virus has been shown to inhibit interferon beta (IFN-$\beta$) production (Boxer et al. 2009). The molecular mechanism of inhibition still needs to be investigated, but it is likely that the C protein blocks the activation of transcription factors which are required to make up the IFN-$\beta$ enhanceosome. In addition, the C protein of different strains of RPV appears to have different effects on the inhibition of IFN-$\beta$, which reflects the action in a strain specific manner. As the C protein proved to be a virulence factor in MV infection (Patterson et al. 2000) and RPV growth (Baron and Barrett 2000), the biological function of the C protein in PPRV is poorly understood and needs to be examined.

### 1.2.3.2 V Protein

The V protein is produced from the P gene through a frame shift, by the incorporation of one G residue during transcription at a particular mRNA editing site. Among morbilliviruses, the editing site (-5'-TTAAAAGGGCACAG-3') is conserved, and in PPRV it is located at 742–756 in the P gene. The length of the V protein is variable among the morbilliviruses: PPRV has 298 amino acids while CDV, RP have 299 amino acids, whereas MV and DMV have 300 and 303 amino acids, respectively (Fig. 1.4a). The predicted molecular mass of the V protein is 32.28 kDa, while the predicted iso-electric point is 4.68. By virtue of having the same initial gene frame, the V protein has an identical N-terminus to the P protein, but after editing and hence a frame shift, the cysteine-rich C-terminus is different between the V and P proteins (Mahapatra et al. 2003). This C-terminus is conserved among the PPRV strains sequenced so far (Fig. 1.4b). This transcriptional editing occurs only in virus-infected cells (Mahapatra et al. 2003).

Like the P protein, the V protein is also phosphorylated, and $\sim 60$ % of the serine residues are revealed to have a high score for phosphorylation as predicted by Netphos 2.0 (Blom et al. 1999). The V protein is shown to be associated with the N and L proteins, which is an indication that the V protein participates in the regulation of viral RNA synthesis (Sweetman et al. 2001). Although the definite roles of V protein is not well established, Tober et al. (1998) found that the lack of a V protein increases viral replication, suggesting its role in the transcription process (Tober et al. 1998). Because these nonstructural proteins are conserved not only in morbilliviruses but also in many other paramyxoviruses, this implies their vital roles in viral growth and pathogenicity in their respective species. Use of site-specific mutations and viruses lacking particular proteins, applying reverse genetics can further clarify their role in the virus life cycle and pathogenicity. A feature of the V protein that has been rather well studied is its role in innate immunity. The majority of paramyxoviruses have the ability to antagonize interferon actions, but this mechanism and the proteins involved are different among viruses. It has been noted that recombinant viruses lacking either the V protein or its cysteine rich

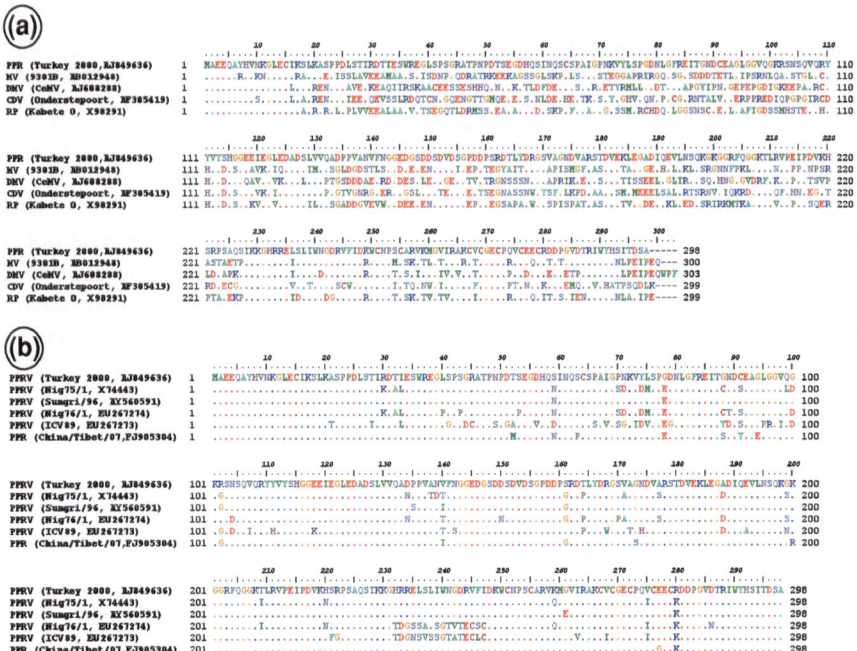

**Fig. 1.4** Alignment of the V proteins. **a** Comparison of the length of the V protein from various morbilliviruses. **b** Comparison of the length and sequence similarity among different PPRV isolates. The amino acids with similar properties were given same color

domain show attenuated growth in vivo, and this is likely due to their ability as an IFN antagonist (Chambers and Takimoto 2009). Studies are required to investigate the functional role of the V protein of PPRV, and its relation to other morbilliviruses despite the fact that PPRV share many common features.

## 1.3 Comparative Genome Analysis

The genome organization of PPRV is identical to that of other members of the genus morbilliviruses, with a slight difference in the genome length. However, all members contain a genome length of less than 16 kb. The genus morbillivirus belongs to the subfamily *Paramyxovirinae* within the family *Paramyxoviridae*. All of the genera within this subfamily share the same genome organization, except the genus *Rubulavirus*, which contains an additional SH (small hydrophobic) gene (Fig. 1.5). The second subfamily *Pneumovirinae* contains two genera named *Metapneumovirus* and *Pneumovirus*. The members of these genera contain a different genome organization compared to the subfamily *Paramyxovirinae*, and

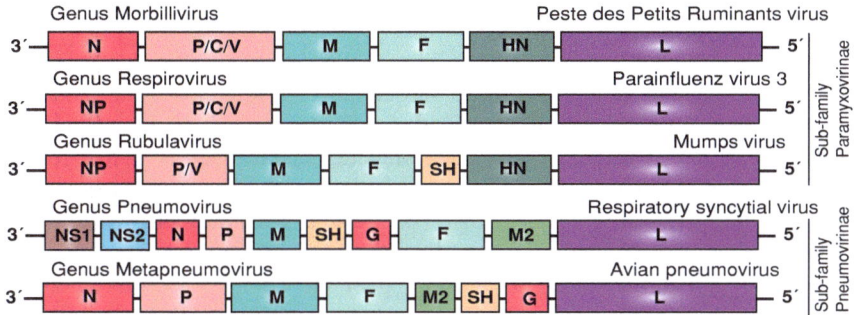

**Fig. 1.5** The genome organization of representative member of each genus of two subfamilies within the family Paramyxoviridae

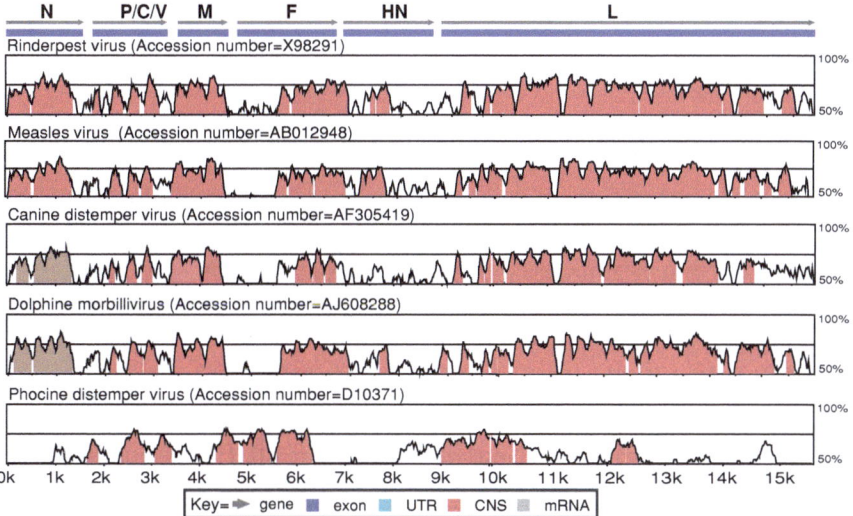

**Fig. 1.6** Global pairwise sequence comparison of PPRV with selected strains of rinderpest, measles virus, canine distemper virus, dolphin morbillivirus, and phocine distemper virus

are schematically presented (Fig. 1.5). Among all of the viruses in the family, *Pneumoviruses* have the most diverse gene order.

Comparison of the complete genome of PPRV to that of other members of genus morbilliviruses indicated that PPRV shown highest nt identity to that of rinderpest strain Kabete 'O' (accession number X98291) followed by MV strain 9301B (accession number AB012948) (Table 1.1). A global pairwise sequence comparison of PPRV to the members of genus morbilliviruses indicated that PPRV is closely related to rinderpest and MV, as predicted by the nt identity (Fig. 1.6).

## 1.4 Conclusion

The reverse genetic system has made substantial contributions to the molecular understanding of the role of different viral proteins in the pathobiology of viruses, especially in MV, CDV, and RPV. Unfortunately, unavailability of the reverse genetic system for PPRV is the greatest hurdle in the advancement of research for understanding the nature of the virus. Therefore, a great gap still exists in our understanding of molecular biology and pathogenesis of the virus. Moreover, this system will not only provide understanding of the complex interplay between viruses and the host, but also will increase our knowledge of PPRV replication, virulence, and cell tropism. Therefore, there is a great need to establish an efficient reverse genetic system for PPRV before any advancement in molecular understanding of PPRV is expected.

## References

Bailey D, Banyard A, Dash P, Ozkul A, Barrett T (2005) Full genome sequence of peste des petits ruminants virus, a member of the Morbillivirus genus. Virus Res 110(1–2):119–124

Bailey D, Chard LS, Dash P, Barrett T, Banyard AC (2007) Reverse genetics for peste-des-petits-ruminants virus (PPRV): promoter and protein specificities. Virus Res 126(1–2):250–255

Baron MD, Barrett T (2000) Rinderpest viruses lacking the C and V proteins show specific defects in growth and transcription of viral RNAs. J Virol 74(6):2603–2611

Baron MD, Goatley L, Barrett T (1994) Cloning and sequence analysis of the matrix (M) protein gene of rinderpest virus and evidence for another bovine morbillivirus. Virology 200(1):121–129

Barrett T, Ashley CB, Diallo A (eds) (2006) Molecular biology of the morbilliviruses. In: Rinderpest and Peste des Petits Ruminants Virus Plagues of Large and Small Ruminants, 2nd edn. Elsevier, Academic Press, London

Bellini WJ, Englund G, Rozenblatt S, Arnheiter H, Richardson CD (1985) Measles virus P gene codes for two proteins. J Virol 53(3):908–919

Blom N, Gammeltoft S, Brunak S (1999) Sequence and structure-based prediction of eukaryotic protein phosphorylation sites. J Mol Biol 294(5):1351–1362

Blumberg BM, Crowley JC, Silverman JI, Menonna J, Cook SD, Dowling PC (1988) Measles virus L protein evidences elements of ancestral RNA polymerase. Virology 164(2):487–497

Bodjo SC, Kwiatek O, Diallo A, Albina E, Libeau G (2007) Mapping and structural analysis of B-cell epitopes on the morbillivirus nucleoprotein amino terminus. J gen virol 88(Pt 4):1231–1242

Bodjo SC, Lelenta M, Couacy-Hymann E, Kwiatek O, Albina E, Gargani D, Libeau G, Diallo A (2008) Mapping the Peste des Petits Ruminants virus nucleoprotein: identification of two domains involved in protein self-association. Virus Res 131(1):23–32

Boxer EL, Nanda SK, Baron MD (2009) The rinderpest virus non-structural C protein blocks the induction of type 1 interferon. Virology 385(1):134–142

Brown DD, Collins FM, Duprex WP, Baron MD, Barrett T, Rima BK (2005) 'Rescue' of mini-genomic constructs and viruses by combinations of morbillivirus N, P and L proteins. J gen virol 86(Pt 4):1077–1081

Buckland R, Giraudon P, Wild F (1989) Expression of measles virus nucleoprotein in Escherichia coli: use of deletion mutants to locate the antigenic sites. J gen virol 70(Pt 2):435–441

Bundza A, Afshar A, Dukes TW, Myers DJ, Dulac GC, Becker SA (1988) Experimental peste des petits ruminants (goat plague) in goats and sheep. Can J Vet Res (Revue canadienne de recherche veterinaire) 52(1):46–52

Cartee TL, Megaw AG, Oomens AG, Wertz GW (2003) Identification of a single amino acid change in the human respiratory syncytial virus L protein that affects transcriptional termination. J Virol 77(13):7352–7360

Chambers R, Takimoto T (2009) Antagonism of innate immunity by paramyxovirus accessory proteins. Viruses 1(3):574–593

Chard LS, Bailey DS, Dash P, Banyard AC, Barrett T (2008) Full genome sequences of two virulent strains of peste-des-petits ruminants virus, the Cote d'Ivoire 1989 and Nigeria 1976 strains. Virus Res 136(1–2):192–197

Choi KS, Nah JJ, Ko YJ, Kang SY, Joo YS (2003) Localization of antigenic sites at the amino-terminus of rinderpest virus N protein using deleted N mutants and monoclonal antibody. J Vet Sci 4(2):167–173

Choi KS, Nah JJ, Ko YJ, Kang SY, Yoon KJ, Jo NI (2005) Antigenic and immunogenic investigation of B-cell epitopes in the nucleocapsid protein of peste des petits ruminants virus. Clin Diagn Lab Immunol 12(1):114–121

Ciancanelli MJ, Basler CF (2006) Mutation of YMYL in the Nipah virus matrix protein abrogates budding and alters subcellular localization. J Virol 80(24):12070–12078

Das SC, Baron MD, Barrett T (2000) Recovery and characterization of a chimeric rinderpest virus with the glycoproteins of peste-des-petits-ruminants virus: homologous F and H proteins are required for virus viability. J Virol 74(19):9039–9047

Dechamma HJ, Dighe V, Kumar CA, Singh RP, Jagadish M, Kumar S (2006) Identification of T-helper and linear B epitope in the hypervariable region of nucleocapsid protein of PPRV and its use in the development of specific antibodies to detect viral antigen. Vet Microbiol 118(3–4):201–211

Dhar P, Muthuchelvan D, Sanyal A, Kaul R, Singh RP, Singh RK, Bandyopadhyay SK (2006) Sequence analysis of the haemagglutinin and fusion protein genes of peste-des-petits ruminants vaccine virus of Indian origin. Virus Genes 32(1):71–78

Diallo A (1990) Morbillivirus group: genome organisation and proteins. Vet Microbiol 23(1–4):155–163

Diallo A, Barrett T, Barbron M, Meyer G, Lefevre PC (1994) Cloning of the nucleocapsid protein gene of peste-des-petits-ruminants virus: relationship to other morbilliviruses. J gen virol 75(Pt 1):233–237

Diallo A, Barrett T, Lefevre PC, Taylor WP (1987) Comparison of proteins induced in cells infected with rinderpest and peste des petits ruminants viruses. J gen virol 68(Pt 7):2033–2038

Diallo A, Minet C, Le Goff C, Berhe G, Albina E, Libeau G, Barrett T (2007) The threat of peste des petits ruminants: progress in vaccine development for disease control. Vaccine 25(30):5591–5597

Durojaiye OA, Taylor WP, Smale C (1985) The ultrastructure of peste des petits ruminants virus. J Vet Med Ser B Infect Dis Immunol Food Hyg Vet Public Health (Zentralblatt Fur Veterinarmedizin Reihe B) 32(6):460–465

Flanagan EB, Ball LA, Wertz GW (2000) Moving the glycoprotein gene of vesicular stomatitis virus to promoter-proximal positions accelerates and enhances the protective immune response. J Virol 74(17):7895–7902

Gattiker A, Gasteiger E, Bairoch A (2002) ScanProsite: a reference implementation of a PROSITE scanning tool. Applied bioinformatics 1(2):107–108

Gibbs EP, Taylor WP, Lawman MJ, Bryant J (1979) Classification of peste des petits ruminants virus as the fourth member of the genus Morbillivirus. Intervirology 11(5):268–274

Giraudon P, Jacquier MF, Wild TF (1988) Antigenic analysis of African measles virus field isolates: identification and localisation of one conserved and two variable epitope sites on the NP protein. Virus Res 10(2–3):137–152

Gombart AF, Hirano A, Wong TC (1993) Conformational maturation of measles virus nucleocapsid protein. J Virol 67(7):4133–4141

Hoffman MA, Banerjee AK (2000) Precise mapping of the replication and transcription promoters of human parainfluenza virus type 3. Virology 269(1):201–211

Horikami SM, Smallwood S, Bankamp B, Moyer SA (1994) An amino-proximal domain of the L protein binds to the P protein in the measles virus RNA polymerase complex. Virology 205(2):540–545

Huber M, Cattaneo R, Spielhofer P, Orvell C, Norrby E, Messerli M, Perriard JC, Billeter MA (1991) Measles virus phosphoprotein retains the nucleocapsid protein in the cytoplasm. Virology 185(1):299–308

Johansson K, Bourhis JM, Campanacci V, Cambillau C, Canard B, Longhi S (2003) Crystal structure of the measles virus phosphoprotein domain responsible for the induced folding of the C-terminal domain of the nucleoprotein. J Biol Chem 278(45):44567–44573

Karlin D, Longhi S, Canard B (2002) Substitution of two residues in the measles virus nucleoprotein results in an impaired self-association. Virology 302(2):420–432

Kaushik R, Shaila MS (2004) Cellular casein kinase II-mediated phosphorylation of rinderpest virus P protein is a prerequisite for its role in replication/transcription of the genome. J Gen Virol 85(Pt 3):687–691

Keita D, Servan de Almeida R, Libeau G, Albina E (2008) Identification and mapping of a region on the mRNA of Morbillivirus nucleoprotein susceptible to RNA interference. Antiviral Res 80(2):158–167

Kingston RL, Baase WA, Gay LS (2004) Characterization of nucleocapsid binding by the measles virus and mumps virus phosphoproteins. J Virol 78(16):8630–8640

Kolakofsky D, Pelet T, Garcin D, Hausmann S, Curran J, Roux L (1998) Paramyxovirus RNA synthesis and the requirement for hexamer genome length: the rule of six revisited. J Virol 72(2):891–899

Kozak M (1986) Regulation of protein synthesis in virus-infected animal cells. Adv Virus Res 31:229–292

Laine D, Trescol-Biemont MC, Longhi S, Libeau G, Marie JC, Vidalain PO, Azocar O, Diallo A, Canard B, Rabourdin-Combe C, Valentin H (2003) Measles virus (MV) nucleoprotein binds to a novel cell surface receptor distinct from FcgammaRII via its C-terminal domain: role in MV-induced immunosuppression. J Virol 77(21):11332–11346

Lamb A, Kolakofsky D (2001) Paramyxoviridae: the viruses and their replication. Fields virology, 4th edn. Lippincott Williams and Wilkins, Philadelphia

Langedijk JP, Daus FJ, van Oirschot JT (1997) Sequence and structure alignment of Paramyxoviridae attachment proteins and discovery of enzymatic activity for a morbillivirus hemagglutinin. J Virol 71(8):6155–6167

Liston P, DiFlumeri C, Briedis DJ (1995) Protein interactions entered into by the measles virus P, V, and C proteins. Virus Res 38(2–3):241–259

Mahapatra M, Parida S, Egziabher BG, Diallo A, Barrett T (2003) Sequence analysis of the phosphoprotein gene of peste des petits ruminants (PPR) virus: editing of the gene transcript. Virus Res 96(1–2):85–98

Malur AG, Choudhary SK, De BP, Banerjee AK (2002) Role of a highly conserved NH(2)-terminal domain of the human parainfluenza virus type 3 RNA polymerase. J Virol 76(16):8101–8109

Meyer G, Diallo A (1995) The nucleotide sequence of the fusion protein gene of the peste des petits ruminants virus: the long untranslated region in the 5′-end of the F-protein gene of morbilliviruses seems to be specific to each virus. Virus Res 37(1):23–35

Mioulet V, Barrett T, Baron MD (2001) Scanning mutagenesis identifies critical residues in the rinderpest virus genome promoter. J Gen Virol 82(Pt 12):2905–2911

Moll M, Klenk HD, Maisner A (2002) Importance of the cytoplasmic tails of the measles virus glycoproteins for fusogenic activity and the generation of recombinant measles viruses. J Virol 76(14):7174–7186

Murphy SK, Parks GD (1999) RNA replication for the paramyxovirus simian virus 5 requires an internal repeated (CGNNNN) sequence motif. J Virol 73(1):805–809

Muthuchelvan D, Sanyal A, Singh RP, Hemadri D, Sen A, Sreenivasa BP, Singh RK, Bandyopadhyay SK (2005) Comparative sequence analysis of the large polymerase protein (L) gene of peste-des-petits ruminants (PPR) vaccine virus of Indian origin. Arch Virol 150(12):2467–2481

Muthuchelvan D, Sanyal A, Sreenivasa BP, Saravanan P, Dhar P, Singh RP, Singh RK, Bandyopadhyay SK (2006) Analysis of the matrix protein gene sequence of the Asian lineage of peste-des-petits ruminants vaccine virus. Vet Microbiol 113(1–2):83–87

Norrby E, Sheshberadaran H, McCullough KC, Carpenter WC, Orvell C (1985) Is rinderpest virus the archevirus of the Morbillivirus genus? Intervirology 23(4):228–232

Ohno S, Seki F, Ono N, Yanagi Y (2003) Histidine at position 61 and its adjacent amino acid residues are critical for the ability of SLAM (CD150) to act as a cellular receptor for measles virus. J Gen Vir 84(Pt 9):2381–2388

Patterson JB, Thomas D, Lewicki H, Billeter MA, Oldstone MB (2000) V and C proteins of measles virus function as virulence factors in vivo. Virology 267(1):80–89

Pawar RM, Raj GD, Kumar TM, Raja A, Balachandran C (2008) Effect of siRNA mediated suppression of signaling lymphocyte activation molecule on replication of peste des petits ruminants virus in vitro. Virus Res 136(1–2):118–123

Plemper RK, Hammond AL, Cattaneo R (2001) Measles virus envelope glycoproteins hetero-oligomerize in the endoplasmic reticulum. J Biol Chem 276(47):44239–44246

Poch O, Blumberg BM, Bougueleret L, Tordo N (1990) Sequence comparison of five polymerases (L proteins) of unsegmented negative-strand RNA viruses: theoretical assignment of functional domains. J Gen Virol 71(Pt 5):1153–1162

Rahaman A, Srinivasan N, Shamala N, Shaila MS (2003) The fusion core complex of the peste des petits ruminants virus is a six-helix bundle assembly. Biochemistry 42(4):922–931

Rahaman A, Srinivasan N, Shamala N, Shaila MS (2004) Phosphoprotein of the rinderpest virus forms a tetramer through a coiled coil region important for biological function. A structural insight. J Biol Chem 279(22):23606–23614

Renukaradhya GJ, Sinnathamby G, Seth S, Rajasekhar M, Shaila MS (2002) Mapping of B-cell epitopic sites and delineation of functional domains on the hemagglutinin-neuraminidase protein of peste des petits ruminants virus. Virus Res 90(1–2):171–185

Riedl P, Moll M, Klenk HD, Maisner A (2002) Measles virus matrix protein is not cotransported with the viral glycoproteins but requires virus infection for efficient surface targeting. Virus Res 83(1–2):1–12

Sato H, Masuda M, Miura R, Yoneda M, Kai C (2006) Morbillivirus nucleoprotein possesses a novel nuclear localization signal and a CRM1-independent nuclear export signal. Virology 352(1):121–130

Servan de Almeida R, Keita D, Libeau G, Albina E (2007) Control of ruminant morbillivirus replication by small interfering RNA. J Gen Virol 88(Pt 8):2307–2311

Seth S, Shaila MS (2001) The hemagglutinin-neuraminidase protein of peste des petits ruminants virus is biologically active when transiently expressed in mammalian cells. Virus Res 75(2):169–177

Shiell BJ, Gardner DR, Crameri G, Eaton BT, Michalski WP (2003) Sites of phosphorylation of P and V proteins from Hendra and Nipah viruses: newly emerged members of Paramyxoviridae. Virus Res 92(1):55–65

Sweetman DA, Miskin J, Baron MD (2001) Rinderpest virus C and V proteins interact with the major (L) component of the viral polymerase. Virology 281(2):193–204

Takeda M, Ohno S, Seki F, Nakatsu Y, Tahara M, Yanagi Y (2005) Long untranslated regions of the measles virus M and F genes control virus replication and cytopathogenicity. J Virol 79(22):14346–14354

Tatsuo H, Okuma K, Tanaka K, Ono N, Minagawa H, Takade A, Matsuura Y, Yanagi Y (2000) Virus entry is a major determinant of cell tropism of Edmonston and wild-type strains of measles virus as revealed by vesicular stomatitis virus pseudotypes bearing their envelope proteins. J Virol 74(9):4139–4145

Tober C, Seufert M, Schneider H, Billeter MA, Johnston IC, Niewiesk S, ter Meulen V, Schneider-Schaulies S (1998) Expression of measles virus V protein is associated with pathogenicity and control of viral RNA synthesis. J Virol 72(10):8124–8132

Varsanyi TM, Utter G, Norrby E (1984) Purification, morphology and antigenic characterization of measles virus envelope components. J Gen Virol 65(Pt 2):355–366

Vongpunsawad S, Oezgun N, Braun W, Cattaneo R (2004) Selectively receptor-blind measles viruses: Identification of residues necessary for SLAM- or CD46-induced fusion and their localization on a new hemagglutinin structural model. J Virol 78(1):302–313

Watanabe M, Hirano A, Stenglein S, Nelson J, Thomas G, Wong TC (1995) Engineered serine protease inhibitor prevents furin-catalyzed activation of the fusion glycoprotein and production of infectious measles virus. J Virol 69(5):3206–3210

Woo PC, Lau SK, Wong BH, Fan RY, Wong AY, Zhang AJ, Wu Y, Choi GK, Li KS, Hui J, Wang M, Zheng BJ, Chan KH, Yuen KY (2012) Feline morbillivirus, a previously undescribed paramyxovirus associated with tubulointerstitial nephritis in domestic cats. Proc Nat Acad Sci U S A 109(14):5435–5440

Yoneda M, Bandyopadhyay SK, Shiotani M, Fujita K, Nuntaprasert A, Miura R, Baron MD, Barrett T, Kai C (2002) Rinderpest virus H protein: role in determining host range in rabbits. J Gen Virol 83(Pt 6):1457–1463

Yoneda M, Miura R, Barrett T, Tsukiyama-Kohara K, Kai C (2004) Rinderpest virus phosphoprotein gene is a major determinant of species-specific pathogenicity. J Virol 78(12):6676–6681

Zhang X, Glendening C, Linke H, Parks CL, Brooks C, Udem SA, Oglesbee M (2002) Identification and characterization of a regulatory domain on the carboxyl terminus of the measles virus nucleocapsid protein. J Virol 76(17):8737–8746

# Chapter 2
# Replication and Virulence Determinants of Peste des Petits Ruminants Virus

**Abstract** The first interaction of the host and pathogen is initiated by receptor binding, which is mediated by the hemagglutination-neuraminidase (HN) protein of Peste des Petits Ruminants Virus (PPRV) and sialic acid on the host cell membrane. A siRNA-mediated study has confirmed that signal lymphocyte activating molecules (SLAM) could be a putative co-receptor for PPRV. As in all paramyxoviruses, RNA-dependent-RNA polymerase (RdRp) binds to the genome promoter, which is a stretch of the nucleotides before the nucleocapsid open reading frame that initiates transcription in a "stop-start" fashion with the contribution of other viral proteins such as matrix, nucleocapsid, and phospho-protein. Viral budding occurs through the neuraminidase activity, which cleaves sialic acid residues from the carbohydrate moieties of glycoproteins. Some of these steps in the replication of PPRV are not fully defined yet. However, among morbilliviruses, PPRV is unique in which the HN protein performs both hemag-glutination and neuraminidase actions, so better reflected as an HN protein instead of H protein. The virus propagation and pathogenicity is directly proportional to that of the host's immune response, parasitic infection, the nutritional level of host, and the age of the animal. This chapter highlights recent studies on PPRV repli-cation, transmission, and the factors, both host and non-host, affecting virus propagation in the host.

**Keywords** Virus lifecycle · Replication · Virulence · Host determinants of pathogenesis

M. Munir et al., *Molecular Biology and Pathogenesis of Peste des Petits Ruminants Virus*, SpringerBriefs in Animal Sciences, DOI: 10.1007/978-3-642-31451-3_2, © The Author(s) 2013

## 2.1 Introduction

The first interaction of the host and pathogen is mediated by receptor binding. PPRV interacts with the host cell membrane through the hemagglutinin-neuraminidase (HN) protein and sialic acid receptors. However, other receptors are also likely to exist for PPRV. Our understanding of PPRV replication and transmission is not fully elucidated yet, although there are significant contributions that make the study of PPRV life cycle possible. These findings have also provided bases to establish further studies and make reliable comparisons to other morbilliviruses. A highly valuable tool, reverse genetic system, will be needed to make the virus life cycle definite, and to investigate the roles of host and non-host factors in virus replication and virulence. This chapter will provide an overview of PPRV with regard to its replication, transmission, and virulence, while highlighting recent studies that have expanded our knowledge about the molecular biology of PPRV and finally lead to establish bases for its possible control.

## 2.2 PPRV Replication and Life Cycle

The first and the foremost step in viral replication is the virus–host receptor interaction. PPRV interacts though the HN protein with sialic acid on the host cell membrane (linked in $\alpha$ 2–3 linkage), which is suggested by the hemagglutinating activities of the PPRV for pig and chicken erythrocytes (Renukaradhya et al. 2002). Recently, a small interfering RNA (siRNA)-mediated study confirmed that SLAM could be a putative co-receptor for PPRV (Pawar et al. 2008). It was revealed that suppression of the SLAM receptor lead to the reduction of the PPRV titer by $\log_{10}$ 1.09–2.28 fold (Fig. 2.1). Additionally, sheep and goat SLAM expressing monkey CV1 cells showed high susceptibility for PPRV growth, and proved to be a reliable source for virus propagation from pathological samples (Adombi et al. 2011). The property to grow in these cells remained identical for other members of morbilliviruses, such as measles virus (MV), canine distemper virus (CDV), and rinderpest virus (RPV). This virus–host interaction, followed by F protein-mediated fusion, leads to release of the nucleocapsid from the viral envelope (illustrated in Fig. 2.2). The large (L) protein then works as an RdRp and initiates the transcription of messenger RNAs (mRNAs) in the cytoplasm. As in all paramyxoviruses, RdRp binds to the genome promoter, which is a stretch of the nucleotide before the nucleocapsid (N) open reading frame (ORF) that initiates transcription in a "stop-start" fashion. There is a series of transcription attenuations across each gene junction, a natural justification for the protein quantity required for the viral replication. The mRNA for the N protein, being at the N-terminus, is most abundantly transcribed and is required most, while the L protein, which is required only in a small amount, is least transcribed due to its being far away from the genome promoters. A quantitative estimation has been made for MV, where P, M,

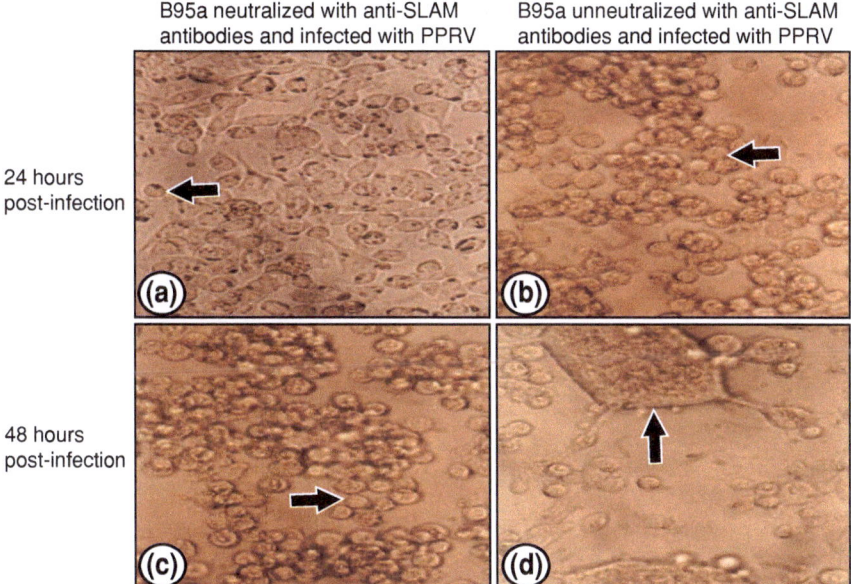

**Fig. 2.1** Inhibition of the SLAM receptor suppresses the replication of PPRV. **a** PPRV infection in the B95a cells for 24 h in which the SLAM neutralized by anti-SLAM antibodies shows no deformities in the cells. **b** PPRV infected for 48 h and non-neutralized SLAM B95a cells shows rounding. **c** PPRV infection in the B95a cells for 24 h in which the SLAM was neutralized by anti-SLAM antibodies shows rounding of the cells. **d** PPRV infected for 48 h and an un-neutralized SLAM B95a cell shows giant cell formation. Note the *arrows* in all the figures (**a–d**). These figures were adapted from Pawar et al. (2008), with permission

F, and H proteins are produced with a percentage of 81, 67, 49, and 39 %, respectively if N protein is taken as 100 % (Horikami and Moyer 1995). Due to the fact that inaccurate transcription occurs in the negative-strand RNA genome, where both mono- and poly-cistronic mRNA are produced, no estimation has been made for the transcription efficacy of each gene in PPRV. As is typical for negative-sense RNA viruses, the RNA produced needs to undergo 5′ capping and 3′ polyadenylation to be efficiently translated by the host ribosomes. These mechanisms have not been well described for the morbilliviruses.

During the course of infection, by an as yet unclear mechanism, transcription switches to replication and produces a full-length antigenome RNA, which is encapsidated by the N protein (Gubbay et al. 2001). Although this transfer of transcriptase to replicase ability of RdRp is complicated, Kolakofsky et al. (2004) have proposed that there are two distinct forms of RdRp in Sendai virus, a member of the same family (Kolakofsky et al. 2004). One form of RdRp is required for transcription while the other acts as a replicase. The contribution of other viral proteins (M, N, and P) in the activity of RdRp cannot be ignored. In MV, the M proteins regulate the function of RdRp but this regulation does not depend on the role of M protein in viral assembly and budding (Suryanarayana et al. 1994;

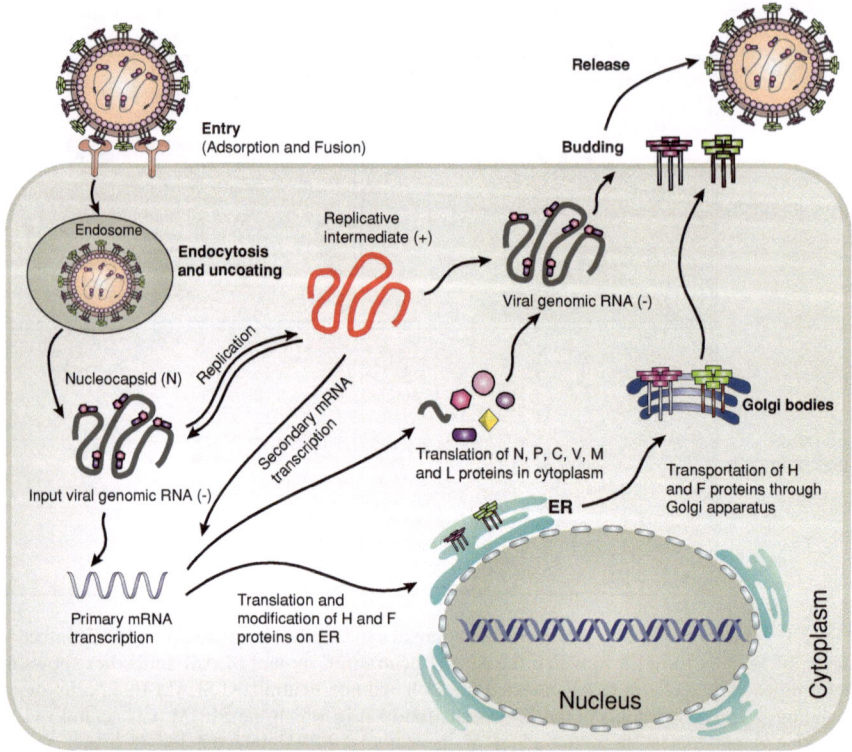

**Fig. 2.2** A generalized morbillivirus replication cycle. Interaction of the viral HN and F proteins with the host plasma membrane leads to viral entry by binding of the HN protein to receptors (SLAM and other unidentified receptors) for PPRV. The rest of the proteins are involved in replication of virus. Briefly, the P protein regulates transcription and replication and assembly of the N protein to nucleocapsids, the M proteins mediate viral assembly, and the HN protein facilitates the budding process when it acts as a neuraminidase in PPRV. Viral genome copies are formed from the minigenome through replicative intermediates. The role of the C and V proteins in PPRV is still not clear. It is believed that these proteins have abilities to abrogate the cellular interferon (IFN-$\alpha/\beta$) responses and hence contribute in the virulence of PPRV

Barrett et al. 2006). In other morbilliviruses, the role of the M protein apart from assembly and budding has not been investigated. Viral budding occurs through neuraminidase activity, which cleaves sialic acid residues from the carbohydrate moieties of glycoproteins (Scheid and Choppin 1974). Among morbilliviruses, PPRV is unique in that the H protein performs both hemagglutination and neuraminidase actions (Seth and Shaila 2001), hence, better reflected as an HN protein instead of an H protein.

The relative levels of the P, V, and C proteins are most likely also regulated in the same way. The editing process can clearly regulate the relative levels of P and V proteins. It is tempting to speculate, based on the functions of these proteins in other morbilliviruses, that during various stages of infection these proteins are expressed at various levels and perform crucial roles in facilitating the virus replication by downregulating the host immunity. However, such functions of these PPRV proteins yet need to be confirmed.

## 2.3 Viral Transmissions and Propagation

### 2.3.1 Non-Host Factors

PPR outbreaks can occur by the close contact of infected and non-infected animals, which are likely to happen in common grazing places. Animals affected by PPR shed the virus in exhaled air, in secretions and excretions (from the mouth, eye and nose, and in feces, semen, and urine) approximately 10 days after the onset of fever. Sneezing and coughing by the infected animal can spread infection, while the transmission between animals in the vicinity can occur through inhalation (over a distance of 10 m) or, unlikely, through inanimate objects (fomite) due to its rapid inactivation in external dry conditions. Spread through ingestion and conjunctival penetration, and by licking of bedding, feed, and water troughs, is also not uncommon. Infection may spread to offspring by feeding them the milk of an infected dam. The exact viral survival in milk has not been demonstrated for PPR, but like rinderpest it is believed that the virus is present in the milk from 1–2 days before the signs appear until 45 days after complete recovery. Like rinderpest, recovered animals show strong immunity and there is no chronic and convalescent carrier state in PPR, but infection is likely to be spread in the subclinical infection during the incubation period. Recently, in an attempt to find out whether the incubatory carriers, as in RPV, could shed PPRV. Couacy-Hymann et al. (2007) confirmed in an experimental infection that infected animals could transmit PPRV before the onset of clinical signs (Couacy-Hymann et al. 2007). The year after, Ezeibe et al. (2008) studied the shedding of virus during the post-recovery state of the animal, and realized that goats infected with PPRV can shed HA virus antigens in feces for 11 weeks after complete recovery (Ezeibe et al. 2008). There is little known about the fragility of PPRV in the external environment. Comparison with rinderpest is likely to be reliable because there are many features in common. Although transmission is not impossible through fomites it is not common either, due to the short life of the virus in dry environment (above 70 °C) and acidic (>5.6) and basic (<9.6) pH. It also cannot resist for a longer time outside the host, due to its short half-life, which is estimated to be 2.2 min at 56 °C and 3.3 h at 37 °C (Rossiter and Taylor 1994).

## 2.3.2  Host Factors

Although cattle, pig, buffalo, and wild ruminants are susceptible to infection, only wild ruminants such as white-tailed deer are fully susceptible and may have a role in the epizootiology of PPR. Little information is available about susceptibility, occurrence, and severity of the disease in wild ungulate species. But a recent report has suggested the role of wildlife in the PPR spread. In this study, Kinne et al. (2010) isolated the virus from different wild small ruminants kept under semi-free-range conditions in the United Arab Emirates (UAE) (Kinne et al. 2010). Sequence analysis of the N gene indicated that the virus belongs to lineage IV, and was different from the viruses already isolated from the Arabian Peninsula (Kinne et al. 2010). Further analysis indicated that these isolates are more closely related to Chinese ones rather to the expected Saudi Arabian isolates. The origin of this new PPRV strain in the region has not been investigated, but it highlights the role of wild ruminants as a possible threat to domestic small ruminants.

Knowledge of the mechanism of PPR virus propagation and dissemination in the host cells is not complete yet. Few studies have demonstrated the sequence of events during virus propagation and its likely ways of spread in the host cell (Scott 1981; Gulyaz and Ozkul 2005). Like other morbilliviruses, PPRV is both lymphotropic and epitheliotropic, and thus the pathological lesions are likely to be severe in organs rich in lymphoid and epithelial tissues (Scott 1981). The PPR virus after invading the host through the respiratory system mainly localizes in the regional lymph nodes (pharyngeal and mandibular) and tonsils, resulting in lymphopenia. The febrile stage may occur on the fifth day and may persist until the sixteenth day post-infection. The resultant viraemia facilitates the dissemination of the virus to all visceral lymph nodes, bone marrow, spleen, and the mucous membranes of both the respiratory and digestive tracts. The virus can be isolated from nasal discharges from the day ninth of virus infection. PPRV then starts multiplying in the gastrointestinal tract, which leads to stomatitis and diarrhea. Scraping from the mucosa of the large intestine and extraction of the mesenteric lymph node can also be used to identify the virus at this moment. An unsuccessful attempt to isolate the virus from the blood of affected animals can be explained by the presence of PPRV-specific neutralizing antibodies that might form a complex with the virus and hence inhibit its recovery. We have recently amplified PPRV nucleic acids directly from filter papers impregnated with the blood of infected animals (Munir et al. 2012), indicating the stability of viral RNA and its presence in the blood. Virus spread to oral lesions has been reported in several studies (Brindha et al. 2001; Gulyaz and Ozkul 2005). Al-Naeem and Abu-Elzein (2008) demonstrated the presence of viral antigen in papules around the oral cavity, which is an indication of the predilection site for viral replication and tropism like the measles virus, a skin lesion-causing virus in humans (Al-Naeem and Abu-Elzein 2008). Although this prediction is helpful to understand the pathogenesis of the disease, further studies are required to confirm that this is not due to other concurrent infections. Bundza et al. (1988) have, for the first time, reported the release of virus particles from the microvilli of intestinal epithelial cells and its shedding in feces (Bundza et al. 1988).

Recently, an unusual staining of PPRV antigen was demonstrated in cortical vessels, proximal tubules, and the epithelium of the renal pelvis. This probably explains the glomerulus filtration of the virus, after pooling in the kidney from the blood stream, and hence secretion in the urine (Kul et al. 2007). A similar localization of other morbilliviruses such as canine- and seal-distemper virus in the urinary system is well established (Kennedy et al. 1989). All morbilliviruses are neurovirulent, and severity depends upon host immunity, specificity of the receptors (such as CD46), and the extent of nervous system infection (Cosby et al. 2002; Kennedy et al. 1989; Yarim and Kabakci 2003). Although this characteristic is not well established in PPRV and RPV, a study conducted by Galbraith et al. (2002) indicates that RPV (Saudi/81) and PPRV (Nigeria 75/1) are neurovirulent when experimentally inoculated into mice (Galbraith et al. 2002). Moreover, a recent study also detected the viral antigen in ependymal cells and meningeal macrophages in natural PPRV infection in 4-month-old sheep (Kul et al. 2007). This interesting feature of PPRV requires further confirmation, because only 1 out of 21 animals showed this sign, but this at least indicates the ability of PPRV to reach cerebrospinal fluid.

## 2.4 Host Determinants of Pathogenesis

To pinpoint the factors required to predispose the animal to infection, it is important to study the epidemiology of the disease and hence its control. Several studies explored factors such as age, sex, breed, and seasons (Amjad et al. 1996; Brindha et al. 2001; Dhar et al. 2002; Munir et al. 2009). Extensive species based antibody surveys have indicated that the level of antibodies against the PPRV N protein was higher in sheep than in goats. Furthermore, it was also observed that goats are more susceptible to infection than sheep in terms of clinical signs. This explains why the virus might have more affinity in goats than sheep. Wosu (1994) has observed that the rate of recovery is lower in goats than in sheep (Wosu 1994). Recently, we have presented a corresponding trend of antibodies in the sheep and goats of governmental livestock farms in Pakistan (Munir et al. 2009). This determinant of pathogenesis needs to be investigated at the molecular level.

The difference in pathogenicity between sheep and goats may not be due to viral affinity, but may be due to a high recovery rate in sheep. In tropical areas, the fertility rate is higher in goats than sheep, which accounts for larger flock replacement by goat offspring. The newborn kids are susceptible to infection after 4 months of age, due to decrease in maternal protective antibodies (Srinivas and Gopal 1996; Ahmed et al. 2005). Waret-Szkuta et al. (2008) recently conducted a serological survey in Ethiopia and declared that age is the main risk factor for the seropositivity in small ruminants (Waret-Szkuta et al. 2008). Bodjo et al. (2006) have suggested the vaccination of the kids and lambs at 75–90 days after birth

(Bodjo et al. 2006). The higher susceptibility in goats may contribute to the severity of PPRV disease in goat populations. It is also true that PPRV infection can spread between goats without affecting sheep in the close vicinity (Animal-Health 2009), but mixed raising of both sheep and goats is considered to be a main risk factor for seropositivity in sheep flocks (Al-Majali et al. 2008). The case fatality rate is also found to be higher in young goats than in adults (Shankar et al. 1998; Atta-ur-Rahman et al. 2004). The sex-biased distribution of antibodies is hard to interpret because of early selling of males and longer maintenance of females.

In subtropical areas, the occurrence of the disease is reported to be more common during winter and rainy seasons (Amjad et al. 1996; Brindha et al. 2001; Dhar et al. 2002). Confinement and restricted movement of the animals, due to rainy seasons in tropical countries, may affect the nutritional status of the animals and hence predispose them to PPRV infection. Some studies have reported major outbreaks in cold and dry weather (Obi et al. 1983; Durojaiye et al. 1983; Opasina 1980), while others reported them soon after the rainy season (Bourdin 1980). This variation is probably explained by the region-dependent differences in animal husbandry conditions and socio-economic status of the farm owner.

## 2.5  Conclusions

Understanding viral replication and the factors influencing it can provide bases for devising the control strategies. Considering that PPRV is a suitable candidate for eradication after RPV, there is a great need to investigate the molecular determinants of PPRV pathogenicity, and to understand the complex interaction between virus and host. Our current knowledge of the virus life cycle shows that both host and environmental factors contribute to the virus transmission and propagation. However, the life of the virus in unusual susceptible hosts such as wildlife and camels remains elusive, but investigation of this could help the efficient planning of animal husbandry and provide a basis for understanding the role of wildlife and camels in the epizootiology of PPRV.

## References

Adombi CM, Lelenta M, Lamien CE, Shamaki D, Koffi YM, Traore A, Silber R, Couacy-Hymann E, Bodjo SC, Djaman JA, Luckins AG, Diallo A (2011) Monkey CV1 cell line expressing the sheep-goat SLAM protein: a highly sensitive cell line for the isolation of peste des petits ruminants virus from pathological specimens. J Virol Methods 173(2):306–313

Ahmed K, Jamal SM, Ali Q, Hussain M (2005) An outbreak of peste des petits ruminants in goat flock in Okara, Pakistan. Pak Vet J 25:146–148

Al-Majali AM, Hussain NO, Amarin NM, Majok AA (2008) Seroprevalence of, and risk factors for, peste des petits ruminants in sheep and goats in Northern Jordan. Prev Vet Med 85(1–2):1–8

Al-Naeem A, Abu-Elzein EM (2008) In situ detection of PPR virus antigen in skin papules around the mouth of sheep experimentally infected with PPR virus. Trop Anim Health Prod 40(4):239–241

Amjad H, Qamar ul I, Forsyth M, Barrett T, Rossiter PB (1996) Peste des petits ruminants in goats in Pakistan. Vet Rec 139(5):118–119

Animal-Health A (2009) Disease strategy: peste des petits ruminants (Version 3.0). Australian veterinary emergency plan (AUSVETPLAN), 3 edn. Primary Industries Ministerial Council, Canberra, ACT

Atta-ur-Rahman, Ashfaque M, Rahman SU, Akhtar M, Ullah S (2004) Peste des petits ruminants antigen in mesenteric lymph nodes of goats slaughtered at D.I. Khan. Pak Vet J 24(3):159–160

Barrett T, Ashley C B, Diallo A (eds) (2006) Molecular biology of the morbilliviruses. In: Rinderpest and peste des petits ruminants virus plagues of large and small ruminants, 2nd edn. Elsevier, Academic Press, London

Bodjo SC, Couacy-Hymann E, Koffi MY, Danho T (2006) Assessment of the duration of maternal antibodies specific to the homologous peste des petits ruminant vaccine "Nigeria 75/1" in Djallonké lambs. Biokemistri 18(2):99–103

Bourdin P (1980) History, epidemiology and economic significance of PPR in West Africa and Nigeria in particular. In: Hill DH (ed) Peste des petite ruminants (PPR) in sheep and goats. In: Proceedings of the international workshop held at IITA Ibadan, Nigeria, 24–26, ILCA (International Livestock Centre for Africa), Addis Ababa, Ethiopia, pp 10–11

Brindha K, Raj GD, Ganesan PI, Thiagarajan V, Nainar AM, Nachimuthu K (2001) Comparison of virus isolation and polymerase chain reaction for diagnosis of peste des petits ruminants. Acta Virol 45(3):169–172

Bundza A, Afshar A, Dukes TW, Myers DJ, Dulac GC, Becker SA (1988) Experimental peste des petits ruminants (goat plague) in goats and sheep. Can J Vet Res (Revue canadienne de recherche veterinaire) 52(1):46–52

Cosby SL, Duprex WP, Hamill LA, Ludlow M, McQuaid S (2002) Approaches in the understanding of morbillivirus neurovirulence. J Neurovirol 8(Suppl 2):85–90

Couacy-Hymann E, Bodjo SC, Danho T, Koffi MY, Libeau G, Diallo A (2007) Early detection of viral excretion from experimentally infected goats with peste-des-petits-ruminants virus. Prev Vet Med 78(1):85–88

Dhar P, Sreenivasa BP, Barrett T, Corteyn M, Singh RP, Bandyopadhyay SK (2002) Recent epidemiology of peste des petits ruminants virus (PPRV). Vet Microbiol 88(2):153–159

Ezeibe MC, Okoroafor ON, Ngene AA, Eze JI, Eze IC, Ugonabo JA (2008) Persistent detection of peste de petits ruminants antigen in the faeces of recovered goats. Trop Anim Health Prod 40(7):517–519

Galbraith SE, McQuaid S, Hamill L, Pullen L, Barrett T, Cosby SL (2002) Rinderpest and peste des petits ruminants viruses exhibit neurovirulence in mice. J Neurovirol 8(1):45–52

Gubbay O, Curran J, Kolakofsky D (2001) Sendai virus genome synthesis and assembly are coupled: a possible mechanism to promote viral RNA polymerase processivity. J Gen Virol 82(Pt 12):2895–2903

Gulyaz V, Ozkul A (2005) Pathogenicity of a local peste des petits ruminants virus isolate in sheep in Turkey. Trop Anim Health Prod 37(7):541–547

Horikami SM, Moyer SA (1995) Structure, transcription, and replication of measles virus. Curr Top Microbiol Immunol 191:35–50

Kennedy S, Smyth JA, Cush PF, Duignan P, Platten M, McCullough SJ, Allan GM (1989) Histopathologic and immunocytochemical studies of distemper in seals. Vet Pathol 26(2):97–103

Kinne J, Kreutzer R, Kreutzer M, Wernery U, Wohlsein P (2010) Peste des petits ruminants in Arabian wildlife. Epidemiol Infect 138(8):1211–1214

Kolakofsky D, Le Mercier P, Iseni F, Garcin D (2004) Viral DNA polymerase scanning and the gymnastics of Sendai virus RNA synthesis. Virology 318(2):463–473

Kul O, Kabakci N, Atmaca HT, Ozkul A (2007) Natural peste des petits ruminants virus infection: novel pathologic findings resembling other morbillivirus infections. Vet Pathol 44(4):479–486

Munir M, Siddique M, Ali Q (2009) Comparative efficacy of standard AGID and precipitinogen inhibition test with monoclonal antibodies based competitive ELISA for the serology of peste des petits ruminants in sheep and goats. Trop Anim Health Prod 41(3):413–420

Munir M, Zohari S, Suluku R, Leblanc N, Kanu S, Sankoh FA, Berg M, Barrie ML, Stahl K (2012) Genetic characterization of peste des petits ruminants virus, sierra leone. Emerg Infect Dis 18(1):193–195

Obi TU, Ojo MO, Durojaiye OA, Kasali OB, Akpavie S, Opasina DB (1983) Peste des petits ruminants (PPR) in goats in Nigeria: clinical, microbiological and pathological features. Zentralblatt fur Veterinarmedizin Reihe B J Vet Med 30(10):751–761

Opasina BA (1980) Epidemiology of PPR in the humid forest and the derived savanna zones. In: Hill DH (ed) Peste des petite ruminants (PPR) in sheep and goats. In: Proceedings of the international workshop held at IITA Ibadan, Nigeria

Pawar RM, Raj GD, Kumar TM, Raja A, Balachandran C (2008) Effect of siRNA mediated suppression of signaling lymphocyte activation molecule on replication of peste des petits ruminants virus in vitro. Virus Res 136(1–2):118–123

Renukaradhya GJ, Sinnathamby G, Seth S, Rajasekhar M, Shaila MS (2002) Mapping of B-cell epitopic sites and delineation of functional domains on the hemagglutinin-neuraminidase protein of peste des petits ruminants virus. Virus Res 90(1–2):171–185

Rossiter PB, Taylor WP (1994) Peste des petits ruminants. In: Coezter JAW (ed) infectious diseases of livestock, vol II. Oxford University Press, Cape Town

Scheid A, Choppin PW (1974) Identification of biological activities of paramyxovirus glycoproteins. Activation of cell fusion, hemolysis, and infectivity of proteolytic cleavage of an inactive precursor protein of Sendai virus. Virology 57(2):475–490

Scott GR (1981) Rinderpest and peste des petits ruminants. In: Gibbs EPJ (ed)Virus diseases of food animals, vol II. Academic Press, London

Seth S, Shaila MS (2001) The hemagglutinin-neuraminidase protein of peste des petits ruminants virus is biologically active when transiently expressed in mammalian cells. Virus Res 75(2):169–177

Shankar H, Gupta VK, Singh N (1998) Occurrence of peste des petits ruminants like diseases in small ruminants in Uttar Pradesh. Indian J Anim Sci 68(1):38–40

Srinivas RP, Gopal T (1996) Peste des petits ruminants (PPR): a new menace to sheep and goats. Livest Advis 21(1):22–26

Suryanarayana K, Baczko K, ter Meulen V, Wagner RR (1994) Transcription inhibition and other properties of matrix proteins expressed by M genes cloned from measles viruses and diseased human brain tissue. J Virol 68(3):1532–1543

Waret-Szkuta A, Roger F, Chavernac D, Yigezu L, Libeau G, Pfeiffer DU, Guitian J (2008) Peste des petits ruminants (PPR) in Ethiopia: analysis of a national serological survey. BMC Vet Res 4:34

Wosu LO (1994) Current status of peste des petits ruminants (PPR) disease in small ruminants. A review article. Stud Res Vet Med 2:83–90

Yarim M, Kabakci N (2003) The comparison of histo-pathological and immunohistochemical findings in natural canine distemper virus infection. Folia Veterinaria 47(2):86–90

# Chapter 3
# Pathophysiology and Clinical Assessment of Peste des Petits Ruminants

**Abstract** Peste des Petits Ruminants (PPR) is a contagious viral infection of both wild and domestic cloven-hoofed small ruminants, characterized by fever, pneumonia, profuse diarrhoea, and inflammation of the mucous membrane of the respiratory and digestive tracts. Depending upon the extent of predisposing factors and the virulence of the virus, PPR severity can be classified as peracute, acute, subacute, and subclinical, but usually PPR follows an acute course of infection. Pathogenesis of PPR starts with the multiplication of the virus in the regional lymph nodes and, after a state of viraemia, the virus disseminates to the surrounding susceptible epithelial tissues. In these tissues, the virus causes observable cytopathic effects that lead to clinical signs and lesions, depending upon the predisposing factors of the host. Zebra striping, developing as a result of severe congestion along the longitudinal folds of the cecum, proximal colon, and rectum, is considered a pathognomonic sign. Despite the viremic state of the disease, the histological changes are more prominent in the oral and intestinal mucosa, where they form multinucleated syncytial cells. Recovered animals show strong immunity, and there is no chronic and convalescent carrier state in PPR. In this chapter, these facts of PPRV are covered comprehensively, and the current literature is reviewed critically.

**Keyword** Pathogenesis · Pathophysiology · Clinical disease · Histopathology · Neurovirulence

## 3.1 Introduction

Peste des Petits Ruminants Virus (PPRV) causes a contagious disease in wild and domestic cloven-hoofed small ruminants. The oral mucous membrane of the respiratory and the digestive tracts are the main susceptible sites for PPRV, as is

M. Munir et al., *Molecular Biology and Pathogenesis of Peste des Petits Ruminants Virus*, SpringerBriefs in Animal Sciences,
DOI: 10.1007/978-3-642-31451-3_3, © The Author(s) 2013

obvious from its descriptive definition: stomatitis pneumoenteritis complex. The histopathological lesions are also predominant in these organs. There are several predisposing factors that determine the virulence of the virus and adversely affect the outcome of the disease. The disease affects young lambs and kids, where it causes severe morality. The neurotropic nature of the PPRV predicts another dimension of this important virus. This chapter will provide an overview of PPR with respect to its clinical features and field diagnosis, while highlighting recent studies that have expanded our knowledge of PPR's clinical picture and pathogenesis.

## 3.2 PPRV Pathogenesis in Small Ruminants

### 3.2.1 Clinical Manifestations

PPR is a contagious viral infection of both wild and domestic cloven-hoofed small ruminants. It is characterized by fever, pneumonia, profuse diarrhoea, and inflammation of the mucous membranes of the respiratory and digestive tracts. Except for frequent pneumonia that could be an interstitial or suppurative and crusty scab around the lips, PPR resembles rinderpest (RP) both clinically and pathologically. Primarily, PPR is a disease of small ruminants where goats are considered more susceptible than sheep, but sheep show higher antibody titers against PPRV and hence better recovery rate than goats. It has been observed that PPR is more severe in West-African goats than in European counterbreeds. Morbidity and mortality rates can reach up to 100 % in severe cases, depending upon the age, breed, body condition, and innate immunity of the host and the virulence of the virus. Moreover, concurrent bacterial and parasitic infections further aggravate the disease.

Large ruminants such as cattle are susceptible and show subclinical infection to PPRV, but they are considered to act as dead-end hosts and are unlikely to participate in the epizootiology of the PPR. Nevertheless, cattle develop humoral immune response against RPV after one infection with PPRV, but this is likely to weaken the immune response to RP vaccination by preventing the replication of the attenuated virus in the RP vaccine (Dardiri et al. 1997; Anderson and McKay 1994). Apart from these, unusual reports of PPR in buffalo (Govindarajan et al. 1997), one-humped camels (Roger et al. 2001), gazelles (Elzein et al. 2004), domestic pigs (Nawathe and Taylor 1979), and American white-tailed deer (*Odocoileus virginianus*) (Hamdy and Dardiri 1976) are also reported.

Depending upon the extent of predisposing factors and the virulence of the virus, PPR severity can be classified as peracute, acute, subacute, or subclinical, but usually PPR runs as an acute course (Braide 1981; Obi et al. 1983; Kulkarni et al. 1996).

### 3.2.1.1 Peracute

Usually, peracute form of disease is observed in kids >4 months old, soon after the depletion of maternal immunity. The disease starts within an incubation period of less than 2 days, which is followed by pyrexia up to 40–42 °C that persists for the next few days. Because of pyrexia, the animal is unable to eat and becomes progressively depressed. During this course, the mucous membrane develops congestion, and occasionally erosion can be seen, which together with oculo-nasal discharges leads to dyspnoea. Constipation in the beginning of this phase converts to profuse watery diarrhea. Animals usually die at the end of this phase, within 4–5 days of pyrexia.

### 3.2.1.2 Acute

In cases where animals survive the peracute phase with non-specific symptoms, they will then go into the more characteristic form of the disease. This form is characterized by serous ocular and nasal discharges after the onset of pyrexia. As the disease progresses, the oculo-nasal discharges become catarrhal, occlude around the nostrils, and predispose to severe dyspnea (Fig. 3.1a, b), sneezing, and coughing followed by restlessness, dry muzzle, and a dull coat (Fig. 3.1c). The incubation period of 3–4 days is accompanied by pyrexia, and 2–3 days post-pyrexia diarrhea or dysentery leads to dehydration and hence emaciation and prostration (Fig. 3.1d). Crusting can develop as a result of congested conjunctiva at the medial canthus, and the conjunctival sac may later fill with thin yellowish fluid, which eventually causes the complete closure of the eyelids. A secondary bacterial infection can further worsen these signs to catarrhal inflammation.

Oral lesions start with rough necrosis on the lower gum below the incisor teeth, and heal rapidly in animals having good prognosis, while for the rest it progresses and covers the dental pad, hard palate, inner side of cheek, dorsal part of the tongue, and around the commissures of the mouth. Animals are reluctant to open their mouth due to pain. Additionally, comparable changes may develop in the mucous membranes of the vulva and the vagina in female animals, and later stages may cause abortion in pregnant animals (Abubakar et al. 2008) (Fig. 3.1e, f). This necrosis is pinpoint grayish with sharply marked foci, which may increase in dimension and amount and turn into shallow non-hemorrhagic erosions. Gentle scrapping of the lesion may produce pale, foul-smelling necrotic material, which consists of dying epithelial cells.

Severe signs of pneumonia such as noisy respiration with extended head and neck, nostril dilation, protruded tongue, and painful cough predispose to poor prognosis. The affected animals then gradually become dehydrated, with sunken eyeballs, and animals often die 10 12 days post pyrexia. The mortality ranges from 70–80 %, while survivors recover after weeks of convalescence.

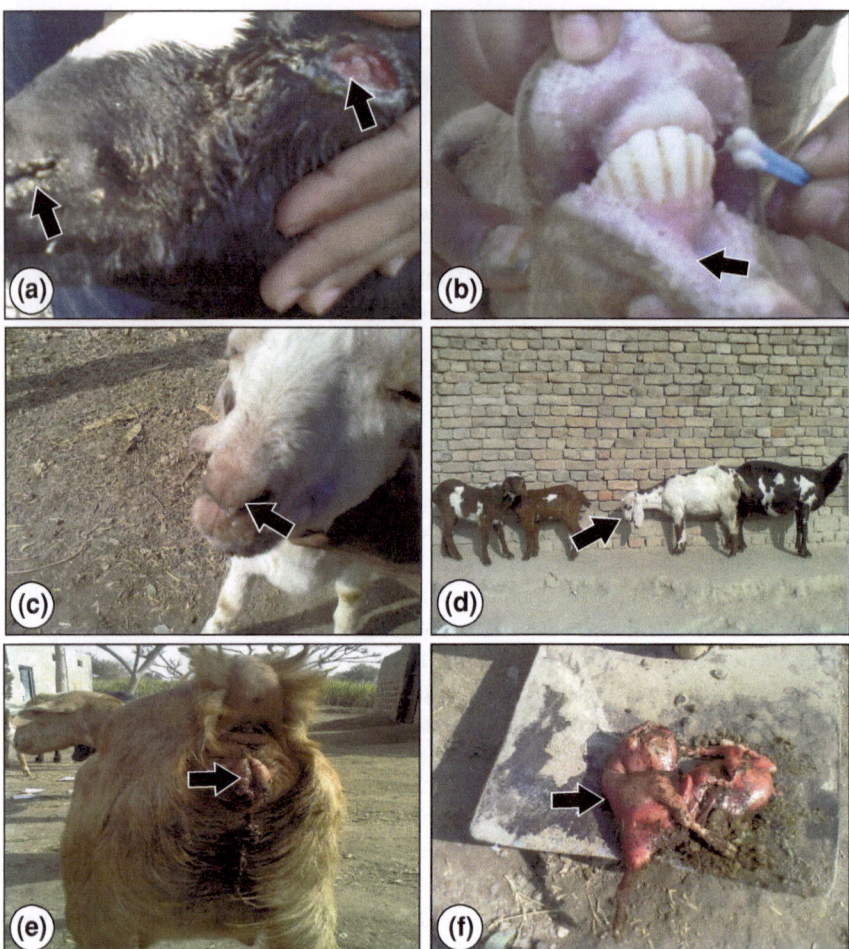

**Fig. 3.1** Clinical picture of animals naturally infected with PPRV. **a** The oculo-nasal discharges become catarrhal with disease progression, occlude around the nostril, and predispose to severe dyspnea. **b** A serous discharge from the oral cavity and crust on the lips (*black arrow*). **c** Coughing, leading to dry muzzle and dull coat. **d** Diarrhea causes dehydration, depression, emaciation, and prostration. **e** The pregnant animals may abort. **f** Aborted fetus from the goat shown in **e**. These images were personally collected from an outbreak of PPRV in Multan district, Pakistan. The samples were confirmed serologically (cELISA) and genetically (real-time PCR) for PPRV

An experimental infection of sheep and goats with live virus infection reveals the pattern of losses of appetite, development of fever, diarrhea, and death (Bundza et al. 1988) (Fig. 3.2).

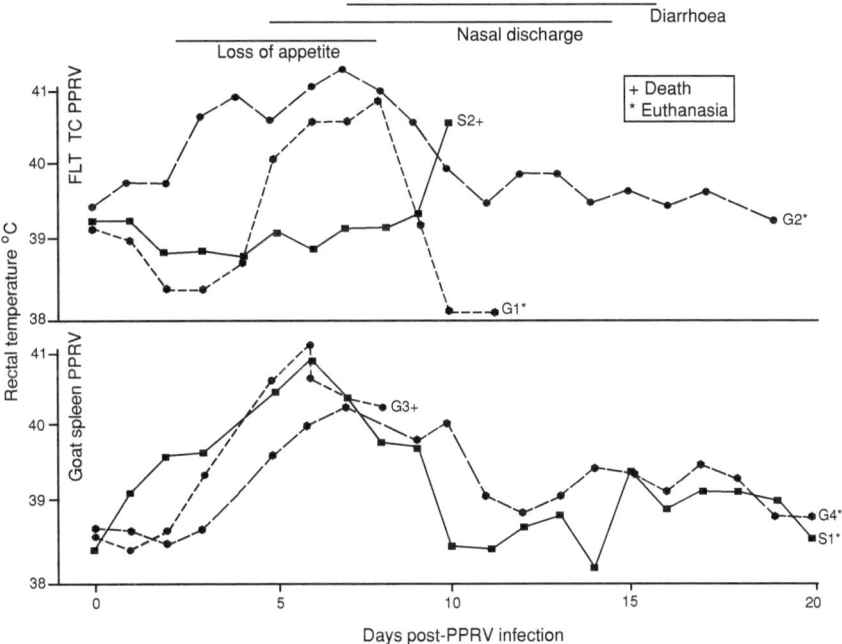

**Fig. 3.2** Clinical assessment and rectal temperature in goats and sheep experimentally infected (intranasally) by a PPRV isolate. This figure was modified from Bundza et al. (1988)

### 3.2.1.3 Subacute

This form of the disease appears in longer incubation period of 6 days. Animals are not severely affected, and lack characteristic signs of PPR, and consequently the mortality rate is very low. Symptoms similar to contagious ecthyma, such as oral crusts due to mucosal discharges, may appear (Diallo 2006). After low-grade pyrexia (39–40 °C), animals usually recover in 10–14 days, but are immunoprotected enough to prevent re-infection and to protect the offspring for at least first 3 months.

### 3.2.1.4 Subclinical

Occasionally, not only sheep and goats but also large ruminants such as buffalo can naturally be infected with the subclinical form of the disease. Animals after this infection are at least testable for the antibodies against PPRV.

## 3.2.2 Pathophysiology

The oral mucous membrane and the respiratory and the digestive tracts are the main susceptible sites for PPRV, as is obvious from its descriptive definition: stomatitis pneumoenteritis complex. Pathogenesis of PPR starts with the multiplication of the virus in the regional lymph nodes, and after a state of viremia the virus disseminates to the surrounding susceptible epithelial tissue. In these tissues, the virus causes observable cytopathic effects that lead to clinical signs and lesions, depending upon the predisposing factors of the host. Congestion of the oral mucosa and the ileo-cecal junction, and occasionally erosions of the oral mucosa, is the only observable lesion in animals that die in the peracute course of the disease. The most common form of the disease is the acute form in which characteristic lesions can be observed in the carcass. The carcass is often emaciated, dehydrated, and the hindquarters are soiled with green to grayish feces. Purulent discharge blocks the nostrils and eyelids, while the lips are hyperemic and encrusted.

## 3.2.3 Gross Lesions

Pathologically, oral lesions are variable from ulcerative to necrotic, which may join together and erode generally on the surface of the oral mucosa, pharynx, upper esophagus, abomasum, and small intestine. Specifically, the dental pad, hard palate, buccal papillae, and the dorsal surface of the tongue are the major affected sites in the oral cavity. Although the lesions are limited to the duodenum, ileum, cecum, and upper colon, occasionally the mucosa of the abomasum may also be affected. In the case of abomasum involvement, congestions, linear engorgement and discoloration of the leaves are evident. Zebra striping, developing as a result of severe congestion along the longitudinal folds of the cecum, proximal colon, and rectum, is considered a pathognomonic sign. The ileo-cecal valve predominantly shows the hemorrhages, but hyperemic, edematous, and ulcerative mucosa can be seen throughout the intestine. Edema and congestion of the lymph nodes, especially the mesenteric, retropharyngeal, and gut-associated lymphoid tissue, are among the common features.

Severe congestion, consolidation, and fibrinous or suppurative pneumonia are most commonly observed lesions on anterior and cardiac lobes of the lungs. Hyperemia, accompanied by frothy exudate, leads to erosions and multifocal ulceration in mucosa of nares and trachea. The skin and heart do not usually show gross lesion but mucopurulent conjunctivitis and swollen spleen with cysts are gross significant features. Rarely, focal degenerative lesions are also noticeable in the liver. Bronchitis, tracheitis, atelectasis, and interstitial pneumonia may be even severe in secondary bacterial infection.

**Table 3.1** Total white blood cell counts, body temperature, and pulse and respiratory rates of lambs and kids with PPRV (reproduced from Cam et al. (2005) with permission)

| Animal | Body temperature (°C) | Pulse rate (bpm)[a] | Respiratory rate (breaths/min) | Total white blood cells ($\times 103/\mu l$)[b] |
|---|---|---|---|---|
| Lambs | | | | |
| 1 | 40.3 | 124 | 56 | 11.6 |
| 2 | 40.4 | 94 | 32 | 7.3 |
| 3 | 40.5 | 136 | 29 | 2.4 |
| 4 | 40.1 | 120 | 34 | 4.5 |
| 5 | 39.5 | 140 | 40 | 2.2 |
| 6 | 40.4 | 100 | 38 | 1.8 |
| 7 | 41.4 | 112 | 68 | 3.2 |
| Kids | | | | |
| 1 | 40.1 | 104 | 30 | 8.8 |
| 2 | 40.5 | 140 | 32 | 5.0 |
| 3 | 39.9 | 128 | 40 | 2.6 |
| 4 | 40.4 | 112 | 68 | 1.3 |

[a] Beats per minute
[b] Reference values for white blood cell counts: Sheep 4 × 103 to 12 × 103/µl, goats 4 × 103 to 13 × 103/µl

## 3.2.4  PPRV in Lambs and Kids

Animals that have recovered from infection, as well as vaccinated animals, induce the production of antibodies in the colostrum. These antibodies are protective for the suckling lambs and kids for at least 3 months. However, this passive immunization is only expected in endemic regions, where either the disease remains at a certain level or vaccination of the herd is practiced. It has been observed that kids and lambs after the age of 3 months are highly susceptible to the infection, more likely due to a decline in passive immunity. Suckling lambs and kids are very much prone to a severe form of the disease, and *"jumping syndrome"* (the author's field experience in which kids and lambs jump, fall, and die) is often observed. Morbidity and mortality rates vary but can be as high as 100 % and 90 %, respectively. These levels are usually lower in endemic areas, and mortality can be as low as 20 % (Taylor et al. 2002).

The disease in kids and lambs has been described in several studies (Aktas et al. 2011; Kul et al. 2008; Taylor et al. 1990; Cam et al. 2005). Collectively, the observations in these studies indicated that PPR can be described in lambs and kids with dullness, anorexia, lacrimation, salivation, and coughing, leading to respiratory stress, diarrhea, and hyperemia in the conjunctiva. The report of Cam et al. (2005) shows the clinical picture of PPRV in lambs (n = 7) and kids (n = 4) (Table 3.1). Common lesions include ulceration and necrosis on the lips, gingiva, buccal mucosa, tongue, and palate. However, the most common and severe lesions are seen around lips and nose.

Infection of PPR along with other concurrent infections leads to severe infection in kids and lambs, which culminates in greater mortality. The concurrent infection of PPRV and pestivirus has been reported to cause stillbirth (Kul et al. 2008). The victim lambs have been characterized grossly and histopathologically. Unique lesions were observed, and are clearly presented and described in the figure legend (Fig. 3.3).

## 3.2.5 Histopathology

Despite the viremic state of the disease, the histological changes are more prominent in the oral and intestinal mucosa, where epithelial cells undergo degenerative changes and tend to join and form eosinophilic inclusion bodies filled with multinucleated syncytial cells (Fig. 3.4a, b, c). Depending on the severity of the disease and the presence of a secondary bacterial infection, the pathological changes vary (Al-Dubaib 2009). In the lungs, multifocal degeneration, ulceration, and necrosis, followed by alveolar type II pneumocytes hyperplasia, which mostly ends up with syncytial cell formation, are prominent features along with others (Aruni et al. 1998; Yener et al. 2004) (Fig. 3.4a). A distinguishing characteristic not found in RP is the presence of multinucleated epithelial giant cells with intranuclear inclusion bodies throughout the lungs. Infiltration of the lymphocytes, plasma cells, and histiocytes into the alveolar septae leads to its hypertrophy and desquamation with intra-alveolar casts. Squamous cell metaplasia with neutrophils infiltration may also be found.

Intestinal mucosal lesions include focal ulceration, edema, and hyperaemia. Some features are noticeable in both naturally and experimentally affected animals, such as atrophy of the villi, reduction of the lymphoid cells in Peyer's patches, dilatation of the cystic crypts of Lieberkuhn with cellular casts, and infiltration of the lamina propria with macrophages and lymphocytes. In cases of liver involvement, narrowing of the sinusoids due to hepatomegaly, and pyknotic nuclei in the necrotic hepatic cells, are obvious with viral inclusions (Fig. 3.4b). Characteristic lesions in the lymph nodes include the hypertrophy of the endo-thelial cells lining the histiocytes, infiltrated sinuses, extensive multifocal necrosis of the trabeculae, and depletion of the lymphocytes. The spleen depicts lesions such as congestion and hyperplasia of the reticulo-endothelial cells, with nuclear inclusion of the macrophages, plasma cells, and giant cells, while acute necrosis of the white pulp of spleen is the main distinguishing feature. The kidneys may undergo necrosis, which is coagulative in nature, and syncytia formation especially in renal tubules (Fig. 3.4d). Hematological studies have indicated the presence of progressive leukopenia and lymphopenia in affected animals, especially in the acute course of the disease.

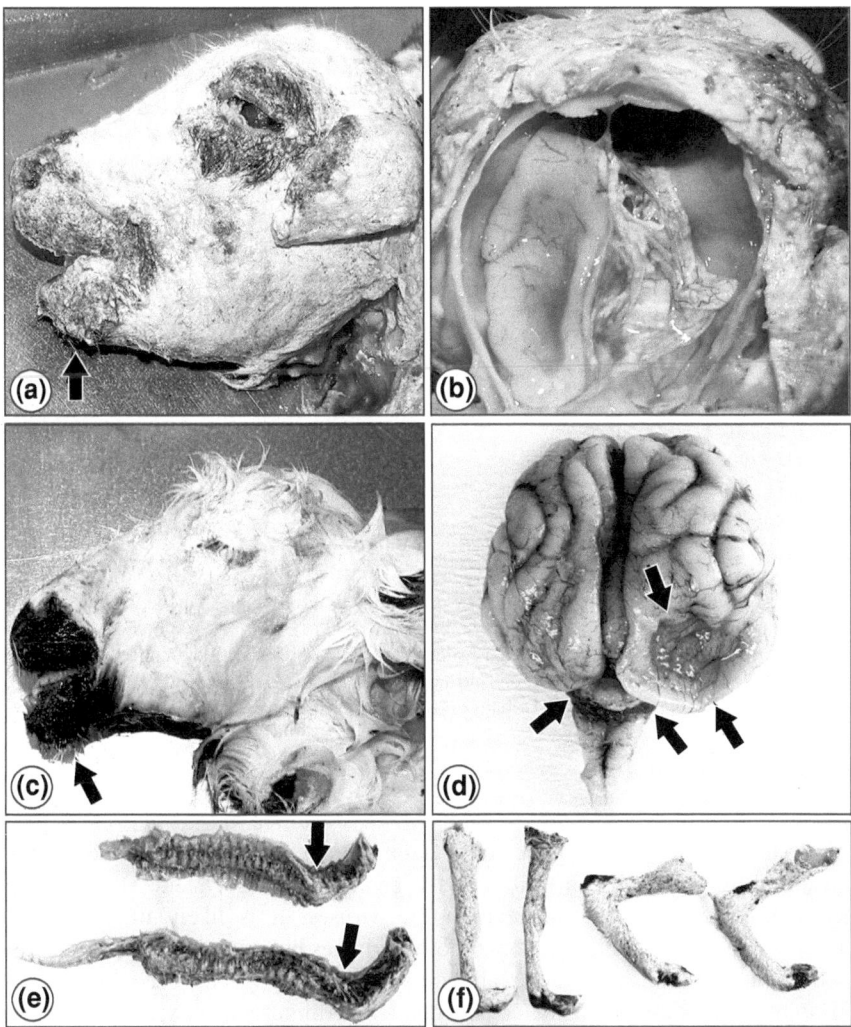

**Fig. 3.3** Ante-mortem and post-mortem lesions in stillborn lambs infected with PPRV and pestivirus. **a** Keratinaceous debris heavily coats the skin. Note the brachygnathia inferior (*black arrow*). **b** Postmortem lesions in the cerebral hemispheres where cortical cells appear as a sac-like layer and the cerebral cortex sticks to the meninges. **c** The fleece is hairy, instead of woolly, and lacks crimp. Note the prognathism inferior (*black arrow*). **d** Observe the defective right parietal-occipital lobe of the cerebrum. **e** Several congenital anomalies were observed in the spinal cord. **f** Abnormalities in the joints such as carpal, tarsal, stifle, and elbow are obvious. Images were adapted from Kul et al. (2008) with permission

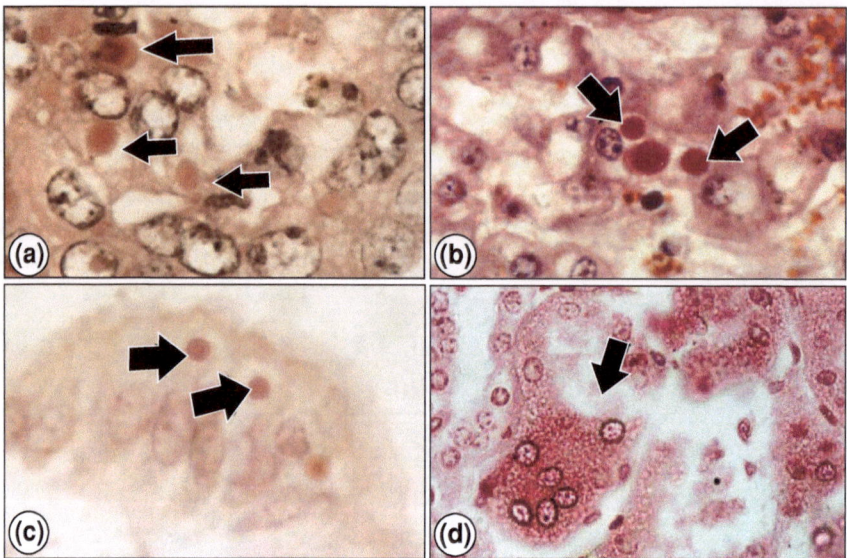

**Fig. 3.4** Histopathological lesions in PPRV infection. **a** Intracytoplasmic inclusion bodies in bronchiolar epithelial tissue under H&E staining at 1000x. **b** *Arrowheads* indicate intracyto-plasmic viral inclusions in hepatocytes. **c** Eroded intestinal epithelial tissue under Macchiavello staining at 400x. **d** Syncytia formation and coagulated necrosis under H&E staining at 250x. These figures are adapted from Al-Dubaib (2009) with permission

## *3.2.6 Neuropathology*

It is clear from the study conducted by Galbraith et al. (2002) that all members of the morbilliviruses cause infection in the central nervous system. However, the neurovirulence may vary between morbilliviruses. It is likely that protective immunity can easily eliminate some morbilliviruses but not others. The protective host immune response and the availability of the virus-specific receptors such as CD46, which is a membrane cofactor protein, determines the entry of the virus into host cells (Galbraith et al. 1998; Galbraith et al. 2002). There is no quantitative information available about the molecular mechanism of entry and pathogenesis of PPRV in the central nervous system. However, based on the available data it is possible to conclude that the Nigeria 75/1 strain of PPRV can cause severe neurovirulence when the viruses are experimentally inoculated into the mouse brain. Furthermore, the virus invasion can be seen in meningeal macrophages and ependymal cells. The presence of PPRV in the ependymal cells clearly provides evidence that PPRV carry the potential to reach and pass the cerebrospinal fluid (Kul et al. 2007). However, such interpretations need to be demonstrated in either natural or experimental infection in the natural host.

It is noteworthy that the presence of PPRV antigens in the central nervous system does not produce any neurological signs either in sheep or in goats.

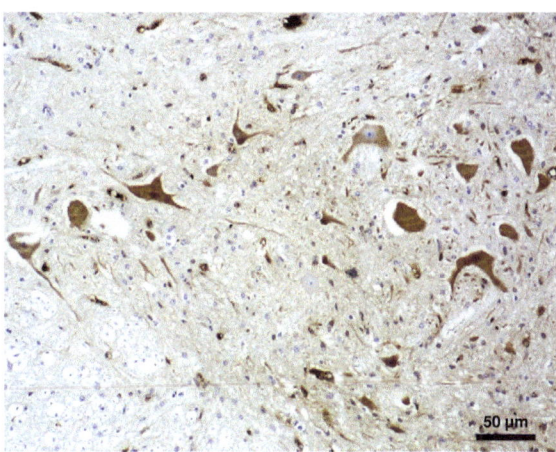

**Fig. 3.5** Immunohistochemical staining of the PPRV antigen in the spinal cord. Staining indicates the antigen presence in the motor neurons and glial cells. This figure is reproduced from Toplu et al. (2011) with permission

The clinical picture of the neurotropic form of PPR in kids and lambs is expected to be severe, as it is obvious in stillborn post-mortem lesions (Kul et al. 2008) (Fig. 3.3). In mice, the infection of PPRV causes disease similar to canine distemper virus, and foci of quite pronounced perivascular inflammation throughout the brain sections are obvious. Further examining the dissemination of PPRV antigen, it has been found that PPRV antigen is detectable in neurons and neuronal processes in the temporal, frontal, and olfactory cortices in both hemispheres, and dendrite processes in the telencephalon layer of the hippocampus (Galbraith et al. 2002; Galbraith et al. 1998). However, oligodendrocytes, astrocytes, microglia, and endothelial cells were consistently negative when PPRV antigen was detected in immunohistochemistry. Results based on these two studies (Galbraith et al. 2002; Galbraith et al. 1998) provided evidence that the presence of a strong immune response in the central nervous system quickly cleared the PPRV infection, or neutralizing antibodies disabled the viral capacity to replicate. Again, this hypothesis still needs to be confirmed in natural hosts, which is not only difficult to do but also expensive. Consistent with the early clearance of PPRV from the central nervous system, it was shown that the presence of other concurrent infections, such as pestivirus, might change the permeability of the blood–brain barriers by damaging the brain tissues, either by endothelial disruption or by direct tissue damage, which allows the PPRV to cross the blood–brain barrier and cause pathogenicity (Toplu et al. 2011). In this study, it was immunohistochemically demonstrated that PPRV antigen could be detected in the motor neurons and glial cells of the spinal cord (Fig. 3.5). Moreover, the staining of the antigen was evident in the neuronal and glial cells of the brainstem, paraventricular areas, and cerebral hemispheres. With these reports, it is now confirmed that PPRV can invade the central nervous system.

## 3.3 Differential Diagnosis

In adults, when PPRV infection appears in a herd for the first time, or when similar diseases are circulating in the region, or disease appears in unnatural hosts, it is likely that PPRV can be confused with any of the following diseases (FAO 2008; Fernandez and White 2010; Rossiter 2004).

### 3.3.1 Rinderpest

Unfortunately, PPRV is a disease of countries where RP was endemic, so that misdiagnosis with PPR is predisposed. However, consideration of a few points makes differential diagnosis easy. RP is mainly a disease of large ruminants (cattle and buffalo) and is now considered to be completely eradicated from the globe, whereas the PPRV is a disease mostly of small ruminants (sheep and goats). However, PPRV can cause subclinical infection, with no obvious clinical signs, in large ruminants, and these animals remain seropositive.

### 3.3.2 Foot-and-Mouth Disease

Foot and mouth disease (FMD) is not a disease of the respiratory system, and therefore animals infected with FMD lack respiratory signs. In PPRV, respiratory signs are very common and obvious. Similarly, diarrhea is often absent in FMD whereas it is usual in PPRV. As the name indicates, FMD virus cause lesions on the feet of infected animals, which are absent in PPRV infected animals. Although both viruses cause severe disease in young lambs, death is more profound and sudden in FMD than in PPRV. The most common feature of both infections is the lesions in the mouth. FMD lesions are very small and do not occlude the oral cavity, and do not cause a foul smell in the affected animals, whereas in PRPV the oral lesions are prominent which can create hindrance in feeding of the animals. The foul smell from the oral cavity of PPRV infected animals is very common. FMD is commonly seen in sheep than goats but vice versa for PPRV.

### 3.3.3 Bluetongue

It is noteworthy that bluetongue (BT) is endemic around the globe whereas PPRV is prevalent in South East Asia, the Middle East, and almost the whole of Africa. Although in both infections pyrexia, lesions in the oral cavity, and discharges are obvious, edema of the head region, bluish discoloration of the oral cavity (especially tongue), coronary banding of the hooves, and the less hairy part of the body are common in BT, which ultimately lead to severe lameness. Involvement of hooves and discoloration of the less hairy part of the body is not reported in PPRV infection. Concurrent infection of BT and PPR in small ruminants complicates the clinical outcome, and requires molecular diagnostic tests to confirm the diseases.

### 3.3.4 Contagious Caprine Pleuropneumonia

Contagious caprine pleuropneumonia (CCPP) and PPRV share some clinical signs, such as difficult breathing and coughing, but lesions in the oral cavity and diarrhea are not present in CCPP. Moreover, CCPP mainly infects goats (sheep are usually not affected) whereas PPRV affects both sheep and goats. As per bacterial infections, lesions in the lungs are more diffuse and the chest cavity is filled with fibrinous fluid, which connects the lungs to the chest wall. In PPRV, severe congestion, consolidation, and fibrinous or suppurative pneumonia are the most commonly observed lesions on the anterior and cardiac lobes of the lung, and a fibrinous connection between the lung and chest cavity is not usually visible.

### 3.3.5 Contagious Ecthyma

Contagious ecthyma (synonyms = orf, sore mouth, contagious pustular dermatitis) and PPRV are usually confused due to the presence of scabs on the lips with contagious ecthyma infection, as with PPRV, which then extends to the mouth and nose and raises an apparent condition identical to PPRV. In contagious ecthyma, oral necrosis, diarrhea, and pneumonia are absent because the alimentary and pulmonary tract are not affected, which are often infected in PPRV infection.

### 3.3.6 Nairobi Sheep Disease

In East Africa, PPR might be confused with Nairobi sheep disease (NSD). In NSD, contrary to PPR, lesions in the oral cavity are either absent or minimal, and goats are really infected. Moreover, NSD is more distributed to the areas where *Phipicephalus appendiculatus* infestation is common.

## 3.3.7  Diarrhea Complex

Coccidiosis or gastro-intestinal helminthic infestations in sheep and goats, and bacterial enteritis caused by *Escherichia coli* and *Salmonella* serovars in kids and lambs, lead to severe diarrhea and can be confused with PPRV. However, typical signs of PPRV such as respiratory involvement and scabs around the oral cavity are lacking.

## 3.3.8  Pneumonic Pasteurellosis

Pneumonic pasteurellosis, as the name indicates, is exclusively a disease of the respiratory tract, which mainly causes pneumonia. The anterior and cardiac lobes of the lungs are packed with dark red spots, which are firm to touch. In normal cases, the alimentary tract is not affected, and therefore oral lesions and diarrhea are lacking, which are otherwise common in PPRV infection. However, oral lesions and diarrhea are not obvious in some cases of PPRV, and result in a complicated situation to differentiate both diseases, and require molecular diagnostic methods such as PCR or cell culturing. The mortality and morbidity rates are higher in PPRV than pasteurellosis. Due to their identical nature, pneumonic pasteurellosis and CCPP have caused the most difficulty in differential diagnosis with PPRV.

## 3.3.9  Heartwater

Heartwater is a disease of almost all subsaharan countries of Africa where *Amblyomma* ticks (a vector necessary for the transmission of *Ehrlichia ruminantium*, a causative agent of heartwater) are prevalent. Like PPRV, heartwater causes both alimentary and respiratory signs, and therefore it is difficult to differentiate until other necessary differential diagnosis means are considered. Heartwater causes clinical disease in both small (sheep, goats) and large ruminants (cattle and buffalo), whereas PPRV causes clinical infection only in small ruminants (sheep, goats). Moreover, in severe cases, heartwater may cause nervous signs that are not manifested in PPRV so far.

## 3.3.10  Mineral Poisoning

Sheep and goats, as per all living creatures, require essential macro- and micro-mineral nutrients in addition to dietary protein, energy, fiber, and water. Balancing these minerals is crucial for all the metabolic activities; deficiency or access of even a single nutrient makes the animals sick. The clinical picture of the animal varies and depends upon the abnormal level of the major elements (calcium, phosphorus, potassium, sodium, chlorine, sulfur, and magnesium) or trace elements (iron, zinc, copper, molybdenum, selenium, iodine, manganese, and cobalt), which may or may not mimic PPRV. Moreover, these ailments are usually curable with mineral supplements.

## 3.4 Conclusions

There have been substantial advances in understanding the clinical assessment of PPRV. Information on the molecular pathogenesis of PPRV is still lacking. The disease is mainly described from naturally infected sheep and goats, and these studies were conducted in the late 1980s. The involvement of concurrent infections in the interpretation of disease investigations cannot be ruled out, especially in the scenarios where PPRV has been reported with BT, contagious caprine pleuropneumonia, and sheep and goat pox. Moreover, owing to the neurotropic nature of PPRV, the pathogenesis of PPRV is not completely clear, which warrants further investigations.

## References

Abubakar M, Ali Q, Khan HA (2008) Prevalence and mortality rate of peste des petitis ruminant (PPR): possible association with abortion in goat. Trop Anim Health Prod 40(5):317–321

Aktas MS, Ozkanlar Y, Simsek N, Temur A, Kalkan Y (2011) Peste des petits ruminants in suckling lambs case report and review of the literature. Isr J Vet Med 66(1):39–44

Al-Dubaib MA (2009) Peste des petitis ruminants morbillivirus infection in lambs and young goats at Qassim region, Saudi Arabia. Trop Anim Health Prod 41(2):217–220

Anderson J, McKay JA (1994) The detection of antibodies against peste des petits ruminants virus in cattle, sheep and goats and the possible implications to rinderpest control programmes. Epidemiol Infect 112(1):225–231

Aruni AW, Lalitha PS, Mohan AC, Chitravelu P, Anbumani SP (1998) Histopathological study of a natural outbreak of peste des petitis ruminants in goats of Tamilnadu. Small Rumin Res 28:233–240

Braide VB (1981) PPR—a review. World Anim Rev 39:25–28

Bundza A, Afshar A, Dukes TW, Myers DJ, Dulac GC, Becker SA (1988) Experimental peste des petits ruminants (goat plague) in goats and sheep. Can J Vet Res (Revue canadienne de recherche veterinaire) 52(1):46–52

Cam Y, Gencay A, Beyaz L, Atalay O, Atasever A, Ozkul A, Kibar M (2005) Peste des petits ruminants in a sheep and goat flock in Kayseri province Turkey. Vet Rec 157(17):523–524

Dardiri AH, DeBoer CJ, Hamdy FM (1997) Response of American Goats and cattle to peste des petits ruminants virus. In: Proceedings of the 19th Annual Meeting of the American Association of Veterinary Laboratory Diagnosticians, Florida, pp 337–344

Diallo A (2006) Control of peste des petits ruminants and poverty alleviation? J Vet Med 53(Suppl 1):11–13

Elzein EM, Housawi FM, Bashareek Y, Gameel AA, Al-Afaleq AI, Anderson E (2004) Severe PPR infection in gazelles kept under semi-free range conditions. J Vet Med 51(2):68–71

FAO (2008) Recognizing Peste des Petits Ruminants, A field manual FAO Corporate document repository. http://www.fao.org/docrep/003/x1703e/x1703e00.htm

Fernandez P, White W (2010) Atlas of transboundary animal diseases. The World Organisation for Animal Health (OIE), Paris

Galbraith SE, McQuaid S, Hamill L, Pullen L, Barrett T, Cosby SL (2002) Rinderpest and peste des petits ruminants viruses exhibit neurovirulence in mice. J Neurovirol 8(1):45–52

Galbraith SE, Tiwari A, Baron MD, Lund BT, Barrett T, Cosby SL (1998) Morbillivirus downregulation of CD46. J Virol 72(12):10292–10297

Govindarajan R, Koteeswaran A, Venugopalan AT, Shyam G, Shaouna S, Shaila MS, Ramachandran S (1997) Isolation of pestes des petits ruminants virus from an outbreak in Indian buffalo (Bubalus bubalis). Vet Rec 141(22):573–574

Hamdy FM, Dardiri AH (1976) Response of white-tailed deer to infection with peste des petits ruminants virus. J Wildl Dis 12(4):516–522

Kul O, Kabakci N, Atmaca HT, Ozkul A (2007) Natural peste des petits ruminants virus infection: novel pathologic findings resembling other morbillivirus infections. Vet Pathol 44(4):479–486

Kul O, Kabakci N, Ozkul A, Kalender H, Atmaca HT (2008) Concurrent peste des petits ruminants virus and pestivirus infection in stillborn twin lambs. Vet Pathol 45(2):191–196

Kulkarni DD, Bhikane AU, Shaila MS, Varalakshmi P, Apte MP, Narladkar BW (1996) Peste des petits ruminants in goats in India. Vet Rec 138(8):187–188

Nawathe DR, Taylor WP (1979) Experimental infection of domestic pigs with the virus of peste des petits ruminants. Trop Anim Health Prod 11(2):120–122

Obi TU, Ojo MO, Durojaiye OA, Kasali OB, Akpavie S, Opasina DB (1983) Peste des petits ruminants (PPR) in goats in Nigeria: clinical, microbiological and pathological features. J Vet Med (Zentralblatt fur Veterinarmedizin Reihe B) 30(10):751–761

Roger F, Yesus MG, Libeau G, Diallo A, Yigezu LM, Yilma T (2001) Detection of antibodies of rinderpest and peste des petits ruminants viruses (Paramyxoviridae, Morbillivirus) during a new epizootic disease in Ethiopian camels (Camelus dromedarius) Revue Méd Vét 152(3):265–268

Rossiter PB (2004) Peste des petits ruminants. In: Coetzer JAW, Tustin RC (ed) Infectious diseases of livestock, vol 2nd. Oxford University Press, Cape Town, Southern Africa, pp 660–672

Taylor WP, al Busaidy S, Barrett T (1990) The epidemiology of peste des petits ruminants in the Sultanate of Oman. Vet Microbiol 22(4):341–352

Taylor WP, Diallo A, Gopalakrishna S, Sreeramalu P, Wilsmore AJ, Nanda YP, Libeau G, Rajasekhar M, Mukhopadhyay AK (2002) Peste des petits ruminants has been widely present in southern India since, if not before, the late 1980s. Prev Vet Med 52(3–4):305–312

Toplu N, Oguzoglu TC, Albayrak H (2011) Dual infection of fetal and neonatal small ruminants with border disease virus and peste des petits ruminants virus (PPRV): neuronal tropism of PPRV as a novel finding. J Comp Pathol 146:289–297

Yener Z, Saglam YS, Temur A, Keles H (2004) Immunohistochemical detection of peste des petits ruminants viral antigens in tissues from cases of naturally occurring pneumonia in goats. Small Rumin Res 51(3):273–277

# Chapter 4
# Immunology and Immunopathogenesis of Peste des Petits Ruminants Virus

**Abstract** Peste des Petits Ruminants Virus (PPRV) is highly immunosuppressive, but the host immune responses against vaccine and infection can mount an effective immunity to PPRV mediated by both cellular and humoral immunity. The T-and B-cell epitopes have been identified against nucleocapsid and hemagglutination-neuraminidase proteins, which provide a foundation for understanding the nature of immunity against PPRV and for the development of assays for epidemiology and surveillance. Based on these results, it is possible to establish the bases for strategies to differentiate infected versus vaccinated animals. There have been a number of reports demonstrating the induction of apoptosis by PPRV, immune suppression caused by PPRV, and cytokine responses against PPRV infection. The hematological and biochemical changes have been described in PPRV-infected animals, either experimentally, or naturally. In this chapter, all of these studies are reviewed and discussed.

**Keywords** Humoral immunity · Cellular immunity · Apoptosis · Immunosuppression · B cell epitope · T cell epitope · Hematology

## 4.1 Introduction

Despite extensive immune-suppression caused by morbilliviruses, including PPRV, infection, and vaccination induce protective immunity to re-infection, and this is protective for the rest of the host's life. Moreover, this protection is independent of the lineage of PPRV and even RPV. There are several reports demonstrating the B- and T cell epitopes in the immunogenic proteins of PPRV; however, the molecular mechanism of humoral and cellular immune induction largely remains elusive. Therefore, the field has remained open to explore which

M. Munir et al., *Molecular Biology and Pathogenesis of Peste des Petits Ruminants Virus*, SpringerBriefs in Animal Sciences, DOI: 10.1007/978-3-642-31451-3_4, © The Author(s) 2013

viral protein(s) cause immunosuppression and what molecular mechanism is involved. Again, lack of a suitable reverse genetic system is a hurdle to proper understanding of this mechanism. There have been several studies explaining the induction of apoptosis, immune suppression, cytokine profiling, and both hematological and biochemical changes in PPRV infected animals, either experimentally or naturally. In this chapter these studies are reviewed and discussed.

## 4.2 Immunity Against PPRV

### 4.2.1 Passive Immunity

Passive immunity is the transfer of readymade antibodies from one individual to another. Maternal passive immunity is one such example, in which maternal antibodies are transferred to the foetus through the placenta resulting in protection of the newborn for a certain period of time. Previous exposure to PPRV infection or development of protective immunity due to vaccines in pregnant dams decides the level of maternal antibodies in the colostrum. The suckling lambs acquire this passive immunity via the colostrum, which is protective for 3–4 months. The level of these maternal antibodies is detectable even until 4 months of age in virus neutralization test but only until the 3rd month in competitive ELISA (Libeau et al. 1992).

A recent study conducted on 23 kids and 26 lambs to ascertain the level of protective maternal immunity demonstrated that lambs and kids are protective for PPRV until age of three and a half to four and a half months (Fig. 4.1), and thus recommended vaccination of the lambs and kids at the ages of 4 and 5 months, respectively (Awa et al. 2003). However, latter studies indicated that vaccination of newborns should be started at the age of 3 months in both kids and lambs in PPRV endemic areas (Bodjo et al. 2006). This discrepancy between the two studies might be due to application of different methods of antibody detection, or use of a virus neutralization test in the former versus cELISA in the latter study. To properly estimate the best time for first vaccination in newborns, it will be ideal to synchronize the pregnancy in sheep and goats so that the immune status of the lambs and kids will be monitored in a harmonized fashioned. Additionally, both virus neutralization test and cELISA should be applied to monitor the neutralizing antibodies and overall antibodies against the virus.

### 4.2.2 Active Immunity

Active immunity is marked by the induction of protective immune responses against invading pathogens, and ultimately results in generation of memory cells, which protect the victims for the rest of their life. Active immunity frequently

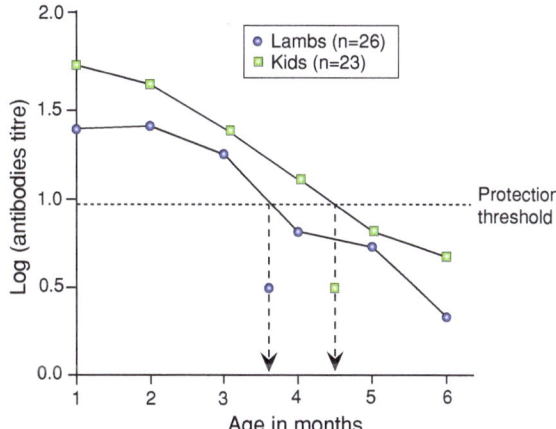

**Fig. 4.1** The level of protective maternal antibodies in newly born lambs and kids over a period of 6 months. This figure was modified from Awa et al. (2003)

comprises both the *cellular* and *humoral* aspects of immunity, and can be acquired by contracting the pathogen (e.g., PPRV infection) or by vaccination (immunization with PPRV vaccine).

### 4.2.2.1  Cellular Immunity to PPRV

Infection and vaccination against PPRV generates both cellular and humoral immunity that are quite effective. Even heterologous vaccination using RPV vaccine renders protection against PPRV, and vice versa. However, in a study using only the glycoproteins H and F of RPV, it was shown that these proteins gave protection against PPRV challenge, but virus-neutralizing antibodies were only found for RPV (Jones et al. 1993). Using a recombinant HN protein of PPRV, Sinnathamby et al. 2001 investigated the immune response to this protein. They showed that immunized goats developed both humoral and cell-mediated immune responses. Furthermore, the generated antibodies could neutralize both PPRV and RPV in vitro (Sinnathamby et al. 2001). On the other hand, it has also been shown that using recombinant vaccinia virus expressing the H or F of RPV could not develop neutralizing antibodies to PPRV, but protection to clinical disease (Romero et al. 1994). All of these studies indicate that the most important immunity is the cellular branch and the neutralizing antibodies may not develop, dependent on vaccination strategy. Another approach to study this was done by Mitra-Kaushik et al. (2001). Using a recombinant nucleocapsid protein (N) of RPV and PPRV, they injected these proteins into mice and studied the antibody and cellular response. They showed that the cellular immune response is both antigen specific and cross reactive between each virus. Further studies in both mice and the natural host identified a conserved T cell epitope (Mitra-Kaushik et al. 2001). This epitope is conserved among all morbilliviruses, explaining the cross protection between RPV and PPRV. Also, the H and HN proteins of RVP and PPRV seem to have quite conserved T cell epitopes (Sinnathamby et al. 2001). These lay in the very conserved N-terminus (amino acids 123–137) and C-terminus (amino acids 242–609).

### 4.2.2.2 Humoral Response to PPRV

Vaccination and infection leads to the development of high quality antibodies. However, as discussed earlier, protection by antibodies seems only to be possible with homologous virus. In our understanding, this means that B cell epitopes are not totally conserved among the morbilliviruses. Some studies have attempted to examine this phenomenon. Choi et al. (2005) have mapped the main B cell epitopes of the N protein. They proposed that the N could be divided into four main antigenic domains, A-I, A-II, C-I, and C-II (Choi et al. 2005). The B cell epitopes of HN have also been determined (Renukaradhya et al. 2002). In this protein monoclonal antibodies bound to two regions: amino acids 263–368 and 538–609. The monoclonal antibodies were tested for their ability to inhibit the neuraminidase activity and hemagglutination activity (Renukaradhya et al. 2002). Of four described monoclonal antibodies, representing the four regions, one cross reacted strongly to RPV (+++), one moderately (++), one weak (+) and one not at all (−). This indicates that the cross reactivity is rather weak, explaining why neutralizing antibodies between PPRV and RPV do not develop.

## 4.3  B- and T-cell Epitopes

Lymphocytes are special leukocytes that play fundamental role in protecting the host against invading pathogens and are crucial in interceding both cell-mediated and antibodies-mediated immune responses. Out of two types of lymphocytes, the T-cells (thymus cells) are involved in cell-mediated immunity. In response to invading pathogen T helper cells, a type of T-cells, get activated, and produce cytokines. Additionally, cytotoxic T-cells, another type of T-cells, produce toxic, and powerful granules. Both secreted cytokines and granules induce the death of pathogen-infected cells (Fig. 4.2). On the other hand, B cells induce the production of antibodies in response to pathogens and these antibodies have strong capacity to neutralize the foreign pathogen (e.g. PPRV). After these direct actions, activated T- and B cells also form memory cells, which remember the specific antigen and mount fast immune response on reintroduction of pathogen (Abbas and Lichtman 2003; von Andrian and Mackay 2000). It is, therefore, essential to map the T- or B cell epitopes (the shortest immunodominant sequence that maintains stimulatory capacity for T- or B-cells) on the viral protein to design efficient recombinant vaccines. Moreover, identification of epitopic domains in the PPRV will help to build foundation for designing methods suitable for the epidemiology and surveillance, and monitoring the immune status of the animals against vaccines so

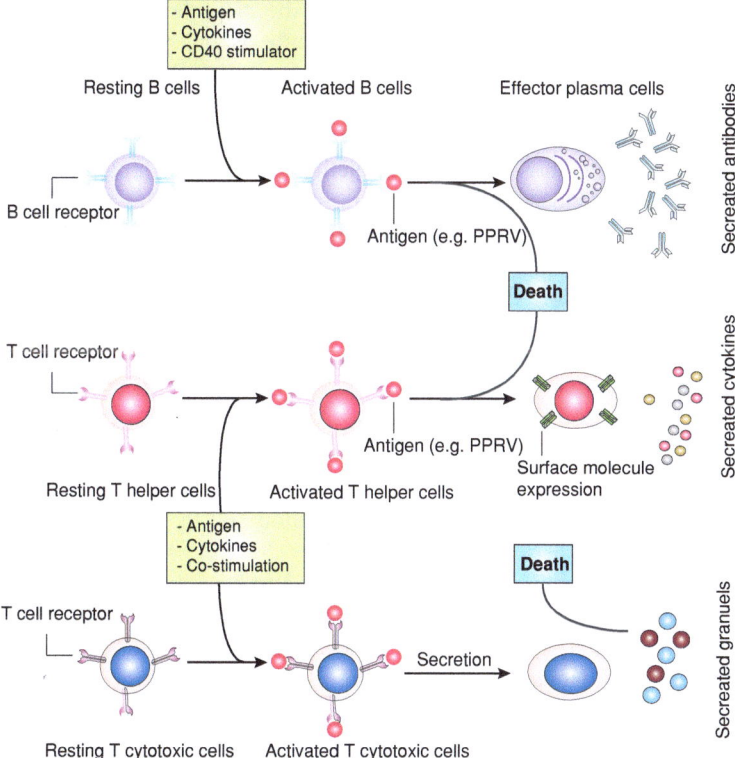

**Fig. 4.2** Activation of B- and T-cells and the mechanism of induction of cell-mediated and humoral immunity

that the disease diagnosis can be made at earliest possible. Such understandings are also fruitful in establishing bases for DIVA (differentiation of infected versus vaccinated animals) strategies. Although immense efforts are still required to map these epitopes in the proteins of PPRV, this part of PPRV research remained most attractive and there have been considerable reports demonstrating the antigenic epitopes on the HN, F, and N proteins of PPRV. Additionally, studies conducted on morbilliviruses such as measles virus, rinderpest virus, canine distemper virus indicated that T-and B-cell epitopes are conserved among morbillivirus. Using these studies as model, it is possible to easily map the proteins of PPRV and will then help to design nonreplicating vector-based subunit vaccines against PPRV.

### 4.3.1  B Cell Epitopes

Being close to the promoter region, the N protein in morbilliviruses is the most abundant protein, and the production of antibodies against the N protein starts early in infection. It is, therefore, likely that the N protein is released into the extracellular compartments during antibody synthesis and binds predominantly to B cell receptors (Laine et al. 2003). Thus, understanding the mechanism and fate of immune responses against the N protein is crucial. In an effort to determine the B cell epitopic profile of the N protein of PPRV, Choi et al. (2005) have characterized the full-length N protein and deletion mutants, by applying monoclonal antibodies (MAbs) and polyclonal antisera in both baculovirus and GST fusions expression systems. Using several MAbs, they have demonstrated the identification of at least four epitopes, designated as A-I, A-II, C–I, and C-II in PPRV strain Nigeria75/1 (Choi et al. 2005). As in other morbilliviruses such as measles virus (Buckland et al. 1989) and rinderpest virus (Choi et al. 2003), A-I and A-II are located in the amino-terminus half from amino acid 1–262, whereas C-I and C-II are located in the carboxyl terminus half from amino acid 448–521. Further analysis using ELISA revealed that epitopes at domains A-II and C-II were immunodominant over epitopic domains at both termini (A-I and C-I). Although the exact location of these epitopes remains to be determined, demonstration of these four domains has provided essential information for the use of N protein in designing serological assays.

The HN protein carrying B-cell epitopes has also been mapped for the presence of immunodominant regions. Using MAbs suggested that the regions from 363 to 368 amino acids and 538–609 amino acids, separated by 171 amino acids, are immunoreactive (Renukaradhya et al. 2002). The MAbs against these regions are not only immunoreactive but are also neutralizing in nature, which provides essential information that these B-cell epitope domains may also participate in the neutralization of the virus. Interestingly, these B-cell epitopes are highly conserved in the PPRV HN protein and it is likely that these appear jointly to present this conserved nature in the tertiary structure of the HN protein.

Almost all morbilliviruses are known to induce cytopathic effects in cells, which makes the presence of the B-cell epitope in the vaccine constructs essential, so that strong neutralization antibodies can be initiated. Such epitopes are also likely to be present on the F protein of PPRV, which demands further investigations.

## 4.3.2 T Cell Epitopes

In order to characterize the T cell epitopes in the PPRV N protein, Mitro-Kaushik et al. (2001) have examined E. coli expressed N protein and determined both antibodies responses and cytotoxic T lymphocyte (CTL) responses in a BALB/c mouse model. They demonstrated that the N protein of both PPRV and RPV induce the class I restricted, antigenic specific, and cross-reactive strong CD8[+] T cell responses. However, the strong antibodies were unable to neutralize the virus. The immunization of mice with purified N protein of PPRV increased the dry weight and there was a significant increase in the proliferation of splenic lymphocytes (Mitra-Kaushik et al. 2001).

Furthermore, the CD4$^+$ cells not only induce the production of virus-specific CTLs but also induce the virus specific B-cells. Such a response in earlier studies left a set of virus-specific memory cells, which may not be the case in PPRV, as it is a cytopathic virus. As is now clear, CD8$^+$ cells play a crucial role in early infection by recognizing nonstructural proteins (C or V in PPRV), and it is likely that the secreted cytokines (IFN-$\gamma$) or MHC-linked cytotoxicity from these cells block the replication of PPRV. Additionally, this could also be associated with the immunosuppression and immunomodulation (Karp et al. 1996). It is, therefore, a fertile area to investigate the correlation between activated CD8$^+$ and major and minor T cell epitopes in the viral proteins in the formation of protective immunity.

Using skin fibroblasts in a proliferation assay presenting MHC class I$^+$ and MHC class II$^-$, it was revealed that CD8$^+$ cells not only respond to the PPRV N protein but also H-2$^d$-restricted CTL epitope, which was further confirmed by direct CTL assay (Mitra-Kaushik et al. 2001). It was also noted that autologous skin fibroblast cells were killed by MHC class I in a restricted fashion when transfected with either PPRV N or RPV N proteins. These properties were concluded to be conserved between PPRV and RPV not only due to the identical functions mentioned above, but also due to the cross-reactivity among infected autologous skin fibroblast cells.

Recently, a bioinformatics supported and experimentally verified report provided evidence for the use of N protein based assays for the DIVA strategies (Dechamma et al. 2006). Based on several evaluation criteria (such as high anti-genic index, and propensity), seven epitopes were chosen in the conserved region of the N protein of PPRV. Only a 19-mer peptide (454–472) was shown to react with antibodies. Infection-immunization studies in rabbit indicated that a 19-mer peptide elicited strong immune responses, and the antibodies remained unaltered with the addition of T-helper antigen. Furthermore, it was identified that the T-helper epitope lies at the amino-terminus, whereas the linear B epitope is located at the carboxyl-terminus region in this 19-mer peptide in the N protein of PPRV. The presence of both epitopes in this short motif in the N protein will help to induce antibodies with greater specificity. These antibodies can be used to differentiate PPRV from RP in ELISA assays.

Besides the N protein, two surface glycoproteins HN and F of morbilliviruses are of immense importance because they induce and confer highly protective immunity. The HN protein of PPRV has been shown to be effective in inducing humoral and cell-mediated immune responses. However, the antigenic sites and the mechanism of these arms of immunity remained unknown until Shaila's group started investigating them. In the first round of investigations, they identified domains in the HN protein of PPRV harboring potential T cell determinants. Out of a highly conserved domain at the amino terminus (113–183 amino acids) of the HN protein of PPRV, H protein of RPV, and measles virus, Sinnathamby et al. (2001) were able to map a 15-mer T cell determinant (from amino acids 123–137). Further investigations provided evidence, although requiring confirmation, that a C-terminal domain (amino acids 242–609) also harboring potential T cell determinant(s) in goats (Sinnathamby et al. 2001).

Latter, the same group used autologous skin fibroblast cells to identify a motif from 400 to 423 amino acids (24 amino acids long) in the HN protein of PPRV, which carries a CTL epitope, and is highly conserved among morbilliviruses, especially in PPRV and RPV (Sinnathamby et al. 2004). This is the only motif identified so far on the HN/H proteins of RPV or PPRV. It was further added that baculovirus expressed H protein of RPV is strongly efficient at inducing neutralizing antibodies, bovine leukocyte antigen (BoLA) class II restricted helper T cell responses and BoLA class I restricted cytotoxic T cell (CTL) responses in cattle immunized against not only H protein of RPV but also against HN protein of PPRV. They have also mapped a BoLA-A11 binding motif (aa 408–416) in the stimulatory domain (Sinnathamby et al. 2004).

## 4.4  Apoptosis Induced by PPRV

Several stimuli, such as viral infection, induce energy dependent cell death, referred to as apoptosis. Certain morphological and biochemical processes are involved in induction of apoptosis including cell shrinkage and partial detachment from the substratum, plasma membrane blebbing, chromatin condensation and intra-nucleosomal cleavage, and ultimately cell fragmentation into apoptotic bodies that are phagocytosed without provoking an inflammatory response (White 1996; Vaux and Strasser 1996).

Induction and inhibition of apoptosis have been described as an advantage for the viruses. It is suggested that viral prevention of apoptosis helps to prevent the premature death of the host cells, and therefore maximizes the chances for viral persistence or increases the virus progeny from infected cells. On the other hands, viruses also induce apoptosis to facilitate the release of progeny viruses and for dissemination to neighboring cells. Additionally, induction of apoptosis helps to establish cytotoxicity that facilitates viral pathogenesis (Roulston et al. 1999).

The nature or the mechanism for the inhibition of apoptosis in PPRV is not defined yet. However, it has been comprehensively studied in other members of the family paramyxoviridae and in the viruses belonging to other families, indicating a universal viral mechanism of self-defense (Laine et al. 2005; Roulston et al. 1999). On the other hand, PPRV-induced apoptosis has been described, which suggests the essential role of this mechanism in viral replication and invasion of host defense mechanisms that limit replication by killing infected cells (Mondal et al. 2001). This study, conducted on goat peripheral blood mononuclear cells, indicated that the induction of apoptosis is directly proportional to that of replication of PPRV (Fig. 4.3a). Additionally, DNA fragmentation, a morphological characteristic of apoptosis, was observed in infected cells. Electron microscopic analysis of the PPRV-infected cells showed margination of chromatin and blebbing of the plasma membrane (Fig. 4.3b). These ultrathin sections also showed formation of apoptotic bodies (Fig. 4.3c), whereas the noninfected cells have showed no such deformities in cells (Fig. 4.3d). All of these points clearly

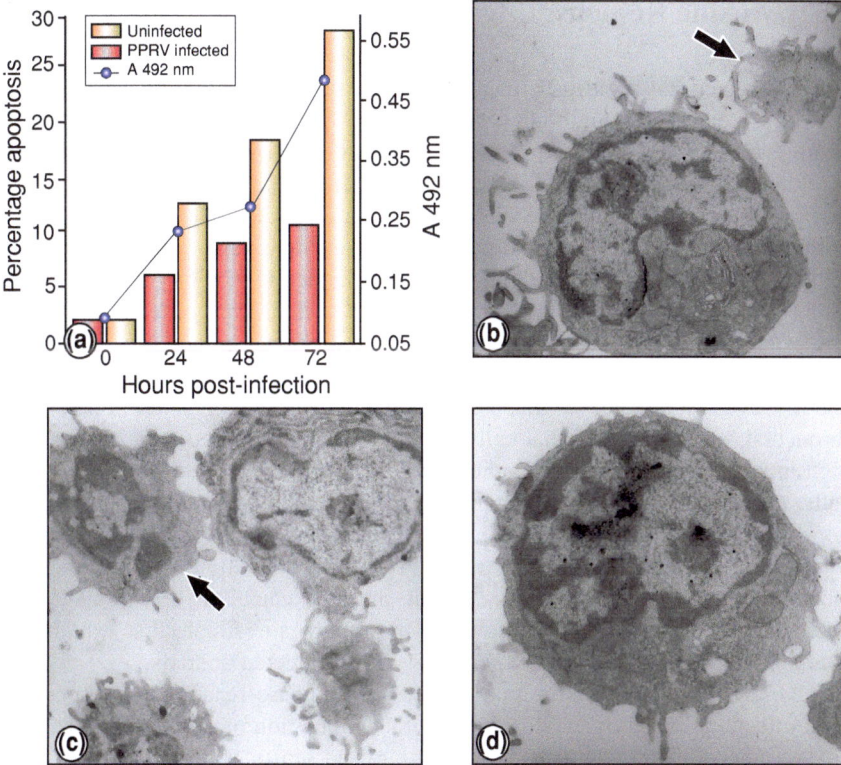

**Fig. 4.3** Apoptosis induction by PPRV virus infection. **a** The level of apoptosis is co-related to the replication of PPRV. **b** Goat cells show deformities in PPRV-infected cells, such as margination of chromatin and blebbing of the plasma membrane. **c** The infected cells showing the formation of apoptotic bodies. **d** Non PPRV-infected normal cells without any deformity. These figures were modified from Mondal et al. (2001) with permission

indicate that PPRV induces apoptosis in at least in goat peripheral blood mono-nuclear cells; however, the molecular mechanism remains elusive. Investigated is required about which viral proteins play a crucial role in this induction and what is the pathway involved in this activation of apoptosis. The nucleoprotein of measles virus, a close relative of PPRV, has been recently identified as an apoptosis inducer (Bhaskar et al. 2011). Due to high similarity of the N proteins between measles and PPRV, it is likely that morbilliviruses share this character in their N protein. It is also suggested that PPRV-induced apoptosis can be associated with viral implication in the immune system. Although the mechanism of immunosuppression in PPRV has not been clearly defined, a character that is believed to be common in all paramyxoviruses, the induction of apoptosis might be associated with the immune suppression, as is practised with measles virus (Schnorr et al. 1997).

## 4.5  Cytokine Responses Against PPRV

All eukaryotes have intrinsic mechanisms to suppress viral replication, which consist of neutralizing antibodies, complements system, and cytokine production. Among other cytokines, interferons (IFNs) are considered to be the principle cytokines involved in antiviral responses. Type II IFN (IFN-$\gamma$) is a potent cytokine that plays a crucial role in the direct inhibition of viral replication and has both immunostimulatory and immunomodulatory effects (Koyama et al. 2008). Naturally, infected animals with PPRV show strong induction of IFN-$\gamma$ in the epithelial lining of the oral cavity, lung and tongue, and a significant difference has been observed when compared to a nonPPRV-infected control group (Atmaca and Kul 2012) (Table 4.1). Immunostaining indicated that capillary mucosa, fibroblasts, and myocytes in the oral submucosa show the highest staining. Additionally, lung (bronchial, bronchial epithelial cells), lingual, and buccal mucosa appeared to be high inducers of IFN-$\gamma$. Besides these organs, intravascular monocytes, syncytial cells, mononuclear cells, and submucosa of the salivary glands were also immunopositive, which indicates the great transmissibility of the PPRV and the ability to induce cytokine to a greater extent. IFN-$\gamma$ plays its antiviral role through the enzyme oligoadenylate synthetase, with the assistance of IFN-$\beta$ and tumor necrosis factor alpha (TNF-$\alpha$). However, no report is available that explains the complex association of these cytokines in PPRV-infected animals.

TNF-$\alpha$ is another cytokine that induces acute phase reactions in response to viral infection. Primarily, TNF-$\alpha$ induces the stimulation of several immune cells, able to induce fever, apoptosis, inflammation, and sepsis, which culminates in the inhibition of viral replication (van Riel et al. 2011). The lungs, interstitial lymphocytes, syncytial cells, and alveolar macrophages of PPRV-infected animals show high expression of TNF-$\alpha$ (Atmaca and Kul 2012). Additionally, submucosal fibroblasts, lungs, and epithelia of salivary gland show significantly high level of TNF-$\alpha$ (Table 4.1). Due to affinity of the PPRV for the epithelial cells, it is likely that TNF-$\alpha$ plays an active role in stimulation of the cell-mediated immune responses which warrants further investigations (Opal and DePalo 2000). Moreover, association of elevated TNF-$\alpha$ and inducible nitric oxide synthetase (iNOS) during PPRV infection may be responsible for the induction of inflammation (Table 4.1). Comparison of TNF-$\alpha$ and IFN-$\gamma$ in measles virus-infected children indicated that IFN-$\gamma$, not TNF-$\alpha$, was significantly different from the noninfected group, which suggests that PPRV may show a difference in its pathogenicity compared to measles virus (Moussallem et al. 2007). Consistent with this, infection of ferrets with canine distemper virus showed no induction of cytokine expression in peripheral blood leukocytes (Svitek and von Messling 2007).

The level of interleukin 4 (IL-4) and IL-10 in PPRV infected animals can be comparatively high in bronchi, bronchial, and interalveolar septum; however, such immunopositivity was not observed to significantly higher from noninfected animals (Table 4.1). IL-4 cytokine is responsible for the inhibition of IFN-$\gamma$

**Table 4.1** Levels of cytokine expression and their statistical analysis

| Cytokine | Tissue | Control group animals | | PPRV positive animals | | Statistical significance ($p > 0.05$) |
|---|---|---|---|---|---|---|
| | | Mean | Standard deviation | Mean | Standard deviation | |
| IFN-Y | Lung | 0.606 | 0.404 | 2.267 | 2.321 | 0.031* |
| | Buccal mucosa | 0.007 | 0.001 | 2.798 | 2.702 | 0.003* |
| | Tongue | 0.006 | 0.001 | 1.461 | 1.198 | 0.003* |
| TNF-α | Lung | 0.03 | 0.0261 | 0.299 | 0.614 | 0.011* |
| | Buccal mucosa | 0.001 | 0 | 0.546 | 0.711 | 0.031* |
| | Tongue | 0.001 | 0 | 0.445 | 0.588 | 0.048 |
| IL-4 | Lung | 0.010 | 0.004 | 0.010 | 0.002 | 0.880 |
| | Buccal mucosa | 0.010 | 0 | 0.024 | 0.059 | 0.880 |
| | Tongue | 0.012 | 0.001 | 0.048 | 0.145 | 0.880 |
| IL-10 | Lung | 0.011 | 0.002 | 0.011 | 0.002 | 0.820 |
| | Buccal mucosa | 0.010 | 0.002 | 0.010 | 0.002 | 0.704 |
| | Tongue | 0.010 | 0.004 | 0.015 | 0.006 | 0.120 |

* $p < 0.05$ was significant

induced monocytes, whereas IL-10 is primarily involved in the suppression of the production of TNF-α and IL-1. This means that high levels of produced TNF-α and IFN-$\gamma$ were not significantly affected by the individual or combined effect of IL4 and IL-10.

A study conducted on PPRV-infected goats explains the distribution and expression of signaling lymphocyte activation molecule receptor (SLAM). SLAM, also known as CD150, is expressed at the surface of T-and B-cells, and are receptors for several morbilliviruses such as measles virus, canine distemper virus, and rinderpest virus. The expression and distribution pattern of SLAM was identical to that of PPRV cell-tropism. Owing to the immunosuppressive nature of PPRV, the mRNA level of SLAM was high in major lymph nodes (mesenteric, hilar, mandibular, and superficial cervical), indicating that PPRV has high affinity for these lymph nodes. The level was also detectable in the respiratory (nasal mucosa) and digestive (duodenum, gallbladder) systems; both of these systems are highly infected with PPRV infection. Additionally, the spleen, thymus, and blood remained highly expressed sites of SLAM receptors under PPRV infection. Despite the fact that PPRV also replicate in the lungs, colon, and rectum, the SLAM receptors were not activated, which partially explains why SLAM is not the major receptor for PPRV infectivity, and that PPRV additionally rely on other receptors for viral pathogenesis (Meng et al. 2011).

## 4.6  Immunosuppression Caused by PPRV

The morbilliviruses, to which PPRV belongs, are highly immunosuppressive. For example, MV induces a profound immunosuppression in small children and infants, which leads to many deaths. This immunosuppression is characterized by lymphopenia, cytokine imbalance leading to impaired cellular immunity, and silencing of peripheral blood lymphocytes (PBL) that, in turn, fail to expand properly (Avota et al. 2010). The major cause of death is, therefore, secondary bacterial infections (Beckford et al. 1985). This can also happen, although not as severely, during vaccination to MV (Griffin and Pan 2009; Hussey and Clements 1996). Nevertheless, after the MV infection or vaccination, individuals have lifelong immunity. In other words, despite severe immunosuppression, an adequate immunity can be induced. The individuals who do not survive MV infection are usually infants with poorly developed immune system.

In PPRV, the level and mechanism of immunosuppression is not completely understood and the available information is meagre. It is most likely that PPRV have some, if not all of these properties. The exact mechanisms are likely to be slightly different from the other morbilliviruses, and also probably among isolates. It has, however, been shown experimentally that the inoculation of highly virulent PPRV (Izatnagar/94) causes immunosuppressive effects in goats (Rajak et al. 2005). In this study it was shown that a virulent PPRV caused leukopenia, lymphopenia, and reduction of antibody response of a specific as well as a nonspecific antigen, indicating immunosuppression (Rajak et al. 2005). These markers of immune suppression were prominent in the acute phase of the disease from 4 to 10 days post-infection, which coincided with severity of clinical symptoms in this study. The lymphotropic nature of PPRV and other morbillivi- ruses renders lymphopenia an important indicator of immune suppression, and it has been shown in both PPRV (Kumar et al. 2001; Raghavendra et al. 1997; Rajak et al. 2005) and RPV (Scott 1981). Interestingly, a vaccine strain was only moderate in all these aspects. It is of note that despite immunosuppression, the immune system can amount an effective immunity to PPRV infection. Although it has been shown that the vaccine strains of PPRV are completely attenuated and cause biologically nonsignificant immune suppression, a mild and transient immunosuppression may lead to predisposition of the animals to secondary infections, which complicates the picture of the disease in the case of an outbreak.

Morbilliviruses, including PPRV, also inhibit the proliferation of human B-lymphoblast cell line (BJAB), in vitro. A study conducted by Heaney et al. (2002) showed that PPRV vaccine strain Nigeria/75/1 causes profound inhibition of freshly isolated, mitogen-stimulated bovine and caprine peripheral blood lymphocytes (PBL) (Fig. 4.4). Moreover, the level of PBL inhibition was found to be more profound in PPRV (50 %) than in a vaccine strain of RPV (30 %) in caprine PBL, especially at a high MOI value of 5. In general, both PPRV and RPV have shown inhibition of PBL proliferation in virus dose dependent ways. Although comparison of wild-type PPRV was not performed in this study, the

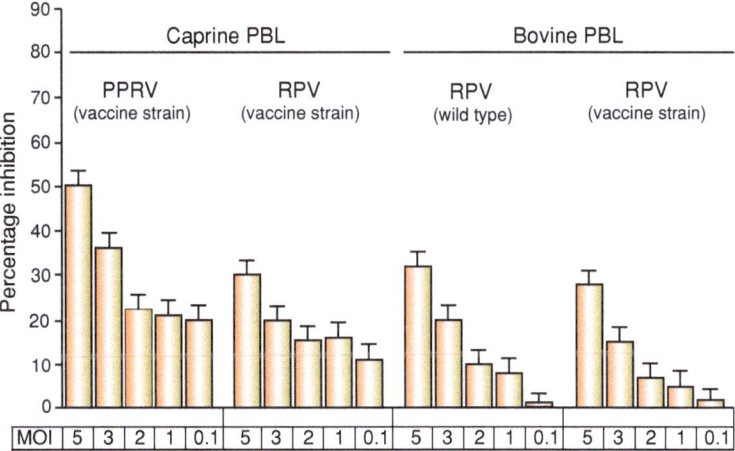

**Fig. 4.4** Inhibition of PBL proliferation by PPRV and RPV infection. (Left panel) PPRV and RPV vaccine strains were cultured with phytohemagglutinin treated caprine PBL for 72 h with a decreasing level of virus MOI from 5 to 0.1. (Right panel) Vaccine strain and wild-type strain of RPV were cultured with phytohemagglutinin-treated bovine PBL for 72 h with a decreasing level of virus MOI from 5 to 0.1. The level of PBL inhibition was estimated with an MTT assay. This figure is modified from Heaney et al. (2002), with permission

level of inhibition remained nonsignificant in wild type and a vaccine strain of RPV in bovine PBL (Heaney et al. 2002). It is, therefore, likely that PPRV act in a manner similar to RPV.

The immunocompromised animals are prone to be severely affected by concurrent infections, and also the disease severity can be many folds higher. It has been recently shown that steroid (dexamethasone) or oxazophorine (cyclophosphamide) induced immune-suppression aggravate the PPR disease in terms of both pathology and dissemination (Jagtap et al. 2012). The drugs induce severe leukopenia and lymphopenia, which allow the PPRV to infect atypical organs such as liver, kidney, and hearts, beside other typical organs. The immunocompromised goats showed viremia for a short time. However, the rate and extent of the disease severity and mortality rate were significantly higher than in noncompromised animals (Fig. 4.5).

In both studies (Jagtap et al. 2012; Rajak et al. 2005), investigating immunosuppression and the fate of immunosuppression in PPRV-infected animals; the specific antibodies against PPRV were not detected. It is tempting to postulate that challenged PPRV interferes in the induction of humoral immune response against antigens, which is further supported by the immune suppression and viremia in the early stage of infection. Since the experiments were limited to the first 10 days of infection, the nature of the immune response against PPRV in the late stages of disease could not be ruled out. However, mortality usually occurs before the seroconversion in the animals. Early detection of PPRV antigens in immunocompromised animals explains the role of these animals in the rapid spread of the

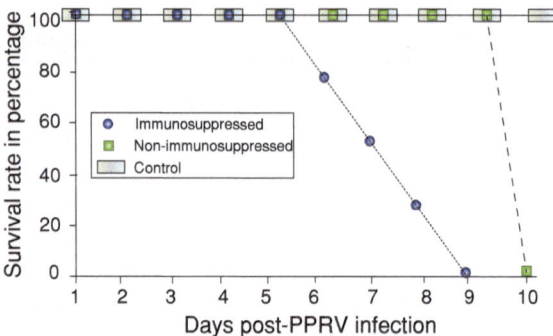

**Fig. 4.5** Survival rate in PPRV infected and immunocompromised or nonimmunocompromised goats. Goats were immunosuppressed by cyclophosphamide and dexamethasone treatments (immunosuppressed) or left untreated (nonimmunosuppressed), followed by PPRV inoculation. The control group consists of goats that were neither immunosuppressed nor PPRV infected. This figure was reproduced from Jagtap et al. (2012) with permission

disease from sick to healthy susceptible animals in the case of an outbreak. In conclusion, the immunosuppressed animals may play a significant role in disease transmission, and can display a severe form of the disease.

The exact mechanism behind MVs ability to suppress immunity is not clearly understood. However, it appears that it is multigenic, and many of the viral genes have been implicated in this property of the viruses (Avota et al. 2010). The nucleoprotein, a soluble variant of it, has been shown to inhibit antibody production (Ravanel et al. 1997), while the glycoproteins, as a complex, impair T cell proliferation (Avota et al. 2010; Niewiesk et al. 1997; Schlender et al. 1996). Furthermore, the V protein of the measles virus has been shown to inhibit IFN-$\alpha/\beta$ and NF-$\kappa$B signaling, culminating in the impaired in the production of IFN-$\alpha/\beta$ (Caignard et al. 2009). In RPV, in contrast to the measles virus, the C protein has been shown to block the induction of type I IFN (Boxer et al. 2009). In that virus, the P protein has also been shown to interact with STAT1 and inhibit IFN signaling. However, the main IFN signaling downregulator is the V protein (Nanda and Baron 2006). In PPRV, the roles of these nonstructural proteins (C and V) are not ruled out in antagonizing immune responses and their contribution to viral pathogenesis. The V protein of PPRV shows high amino acid identity to that of the V protein of measles virus, and therefore it is likely that IFN inhibitory character of PPRV lying in the V protein. Our preliminary results indicate that both termini of the V protein are involved in the inhibition of IFN-$\alpha/\beta$ and NF-$\kappa$B signaling (Munir et al. unpublished data). However, these findings need to be confirmed in both in vivo and in vitro systems.

## 4.7   Hematological Responses Against PPRV

Due to hemorrhages, diarrhea and the affinity of the PPRV for lymphoid organs, it is important to screen the blood components and the preferences of virus over depletion of a specific component. As expected, the hemorrhages in the digestive system and the liver reduce the number of erythrocytes and hematocrit values significantly in kids naturally infected with PPRV (Sahinduran et al. 2012) (Table 4.2). This reduction in hematocrit values could also be due to severe diarrhea caused by PPRV. Although the contribution of stress to the reduction of neutrophils numbers cannot be ruled out, the marked immunosuppression, indicated by leukopenia, monocytes, lymphopenia, was evident. The number of eosinophil remained unaltered because, these immune-cells are primarily associated in parasitic infection. Another study conducted to ascertain the impact of bodyweight, sex, location, and PPRV infection on hematological parameters showed that none of these factors influenced packed cell volume and hemoglobin concentration. On the other hand, do have an effect on neutrophil and lymphocyte (Aikhuomobhogbe and Orheruata 2006).

Thrombocytes or platelets (PLT) are essential components of blood and are responsible for hemostasis, which leads to formation of blood clots in the case of injuries. Activated partial thromboplastin time (APLTT) and prothrombin time (PT) are indicators of the intrinsic and extrinsic pathways of coagulation, respectively. These markers are used to determine the clotting tendency of blood, which indirectly indicates the status of the liver, such as liver damage and vitamin K status. Studies have indicated that PPRV infection causes thrombocytopenia (decrease in thrombocytes) to be significantly higher than in noninfected animals (Table 4.2). Moreover, infection of PPRV in kids also increases the APLTT and PT time, which directly demonstrates one of the possibilities: decreased production of PLT from bone marrow, increased consumption of PLT, loss of PLT due to peripheral destruction, or a combination of these factors. However, the traumatic nature of the liver in PPRV-infected animals indicates that the delayed APLTT and PT are due to trauma and disseminated intravascular coagulation.

Albumin and globulin are the proteins that play crucial roles in blood. Albumin is mainly responsible for the regulation of colloidal osmotic pressure of blood by binding to cations (such as $Ca^+$, $Na^+$ and $K^+$), hormones, bilirubin, and thyroxine (T4). The globulin part of the immune system by acting as antibodies of various classes, and it is involved in combating infection and tissues. Compared to healthy animals, the level of globulin increased significantly in PPRV-infected animals, whereas albumin decreased significantly, which led to an increase in total blood protein and decrease in ratio of albumin and globulin proteins in the blood (Yarim et al. 2006) (Table 4.2).

**Table 4.2** Hematological and biochemical values in PPRV infected and healthy kid groups

|  | Parameters | Infected group (n = 12) | Control group (n = 5) | Level of significance |
|---|---|---|---|---|
| Haematological | Total leukocyte ($\times 10^9$/L) | 2.11 ± 0.29 | 10.68 ± 1.25 | ≤0.001** |
|  | Neutrophil ($\times 10^9$/L) | 9.17 ± 0.38 | 1.95 ± 0.43 | ≤0.001** |
|  | Lymphocyte ($\times 10^9$/L) | 1.88 ± 0.25 | 7.70 ± 0.57 | ≤0.001** |
|  | Erythrocyte ($\times 10^{12}$/L) | 3.29 ± 0.23 | 7.89 ± 0.25 | ≤0.001** |
|  | Monocyte ($\times 10^3$/µL) | 1.4 ± 0.1 | 1.2 ± 0.3 | ≤0.001** |
|  | Eosinophil ($\times 10^3$/µL) | 0.3 ± 0.03 | 0.4 ± 0.03 | >0.05 |
|  | Total protein (g/dL) | 7.2 ± 0.3 | 6.8 ± 0.5 | ≤0.05* |
|  | Albumin (g/dL) | 2.3 ± 0.2 | 2.7 ± 0.4 | ≤0.001** |
|  | Globulin (g/dL) | 4.9 ± 0.4 | 4.2 ± 0.8 | ≤0.001** |
|  | Albumin/globulin ratio | 0.48 ± 0.07 | 0.68 ± 0.24 | ≤0.001** |
|  | Hemoglobin (g/dL) | 97.71 ± 4.64 | 82.20 ± 1.79 | >0.05 |
|  | Hematocrit( %) | 17.14 ± 1.22 | 29.85 ± 1.75 | ≤0.001** |
|  | PT (s) | 18.65 ± 0.42 | 11.26 ± 0.31 | ≤0.001** |
|  | APTT (s) | 34.76 ± 0.63 | 30.36 ± 0.67 | ≤0.01* |
|  | PLT ($\times 10^{11}$/L) | 2.04 ± 0.02 | 5.18 ± 0.23 | ≤0.001** |
| Biochemical | BUN (mg/dL) | 30.75 ± 9.39 | 13.36 ± 0.84 | ≤0.01** |
|  | Creatinine (mg/dL) | 2.67 ± 0.11 | 1.49 ± 0.10 | ≤0.001** |
|  | ALP (U/L) | 449.00 ± 47.90 | 181.64 ± 42.75 | ≤0.01* |
|  | AST (U/L) | 432.00 ± 14.52 | 181.80 ± 30.74 | ≤0.001** |
|  | ALT (U/L) | 47.08 ± 1.98 | 30.79 ± 1.64 | ≤0.001** |
|  | GGT (U/L) | 141.58 ± 51.82 | 39.88 ± 5.25 | >0.05 |
|  | Total bilirubin (mg/dL) | 0.33 ± 0.12 | 0.22 ± 0.05 | ≤0.05* |
|  | Direct bilirubin (mg/dL) | 0.23 ± 0.08 | 0.16 ± 0.04 | ≤0.05* |
|  | Indirect bilirubin (mg/dL) | 0.10 ± 0.05 | 0.05 ± 0.02 | ≤0.05* |
|  | Cholesterol (mg/dL) | 108.1 ± 11.3 | 106.6 ± 14.3 | >0.05 |
|  | Serum sialic acid | 82 ± 8.9 | 62.2 ± 3.8 | 0.05* |

*Moderately significant, **highly significant, non-significant $p > 0.05$. This table was generated from Yarim et al. (2006) and Sahinduran et al. (2012)

## 4.8  Biochemical Responses to PPRV

Urea is a protein by-product in liver that is removed from blood via the kidneys. The measurement of nitrogen in the blood in the form of urea is considered to be indicative of renal function. Creatinine is a by-product of creatine phosphate in muscle, and is primarily filtered out from the blood by the kidneys. Infection of kids with PPRV significantly induces the production of both blood urea nitrogen and creatinine, compared to untreated kids, which indicates the replication of PPRV in these organs to initiate deterioration (Sahinduran et al. 2012). Four enzymes, aspartate aminotransferase (AST), alanine aminotransferase (ALT), alkaline phosphatase, and gamma glutamyl-transferase (GGT), are considered to be markers for liver functionality. Two independent studies (Yarim et al. 2006; Sahinduran et al. 2012) demonstrated that PPRV-infected animals show significantly elevated levels of all of these enzymes except GGT, which was found significantly different in one study (Table 4.2).

Bilirubin is a breakdown product of hemoglobin catabolism, and is secreted in bile (animal faeces) and urine, which gives the yellow color to urine (due to its by-product urobilin) and the brown color to faeces (due to its by-product sterocobilin). These colors are makers for the presence of abnormal bilirubin that might occur due to the presence of disease. Bilirubin is processed first in the spleen (direct bilirubin) and then in the liver (indirect bilirubin). Infected animals show significantly elevated levels of both direct and indirect bilirubin, and hence the total bilirubin in the blood serum compared to healthy animals. However, cholesterol level (another indicator of devastating diseases) remained unaffected in PPRV infected or noninfected animals (Table 4.2).

Sialic acid is an integral component of cell membranes in both animal and plant cells, and which acts as a receptor for some members of the paramyxoviruses (to which PPRV belong) and orthomyxoviruses (such as influenza viruses). The level of sialic acid in the serum is associated with disease, causing destruction in liver, and in cancers. This serum sialic acid also acts as a marker in acute phase disease reactants, especially those containing sialic acid residue in the oligosaccharide side chain. Infection of animals with PPRV elevated the level of sialic acid in the serum compared to healthy noninfected animals (Yarim et al. 2006) (Table 4.2). This level also correlates with the liver functions test, and gives strong clues that PPRV cause damage to the liver; however, cell-mediated immune responses and acute phase reactions in PPRV can also lead to increase in sialic acid in serum. Regardless of the cause of this induction, serum sialic acid can be used as a marker for the diagnosis of PPRV infection in small ruminants.

## 4.9   Conclusions

The induction of protective immunity and immunosuppression at the same time in PPRV-infected animals is an interesting phenomenon, which probably determines the clearance of the virus from the host. The preference for the cell-mediated and humoral immunity over the clearance of virus infection and mounting protective immunity is not understood, which leaves open many possibilities to explore. In the light of the findings that the HN and N proteins harbor T cell and/or B cell epitopes, it is of primary importance to identify alternative mechanisms that can be used to prime the small ruminants against the N protein of PPRV, which might provide a major target for virus-specific cell-mediated immunity. Moreover, simulation of similar responses by use of IFN-$\gamma$ and other cytokines will also help to elucidate the mechanisms associated with morbilliviruses in general and PPRV in specific immunosuppression and immunomodulation.

# References

Abbas AK, Lichtman AH (2003) Cellular and molecular immunology, 5th edn. Saunders, Philadelphia

Aikhuomobhogbe PU, Orheruata AM (2006) Haematological and blood biochemical indices of West African dwarf goats vaccinated against Pestes des petit ruminants (PPR). African J Biotechnol 5(9):743–748

Atmaca HT, Kul O (2012) Examination of epithelial tissue cytokine response to natural peste des petits ruminants virus (PPRV) infection in sheep and goats by immunohistochemistry. Histol Histopathol 27(1):69–78

Avota E, Gassert E, Schneider-Schaulies S (2010) Measles virus-induced immunosuppression: from effectors to mechanisms. Med Microbiol Immunol 199(3):227–237

Awa DN, Ngagnou A, Tefiang E, Yaya D, Njoya A (2003) Post vaccination and colostral Peste des petits ruminants antibody dynamics in research flocks of Kirdi goats and Fulbe sheep of North Cameroon. In: Jamin JY, Seiny Boukar L, Floret C (eds) Savanes africaines: des espaces en mutation, des acteurs face à de nouveaux défis. Actes du colloque, Garoua, Cameroun. Prasac, N'Djamena, Tchad—Cirad, Montpellier, France

Beckford AP, Kaschula RO, Stephen C (1985) Factors associated with fatal cases of measles. A retrospective autopsy study. South African Med J Suid-Afrikaanse tydskrif vir geneeskunde 68(12):858–863

Bhaskar A, Bala J, Varshney A, Yadava P (2011) Expression of measles virus nucleoprotein induces apoptosis and modulates diverse functional proteins in cultured mammalian cells. PLoS One 6(4):e18765

Bodjo SC, Couacy-Hymann E, Koffi MY, Danho T (2006) Assessment of the duration of maternal antibodies specific to the homologous peste des petits ruminant vaccine "Nigeria 75/1" in Djallonké lambs. Biokemistri 18(2):99–103

Boxer EL, Nanda SK, Baron MD (2009) The rinderpest virus non-structural C protein blocks the induction of type 1 interferon. Virology 385(1):134–142

Buckland R, Giraudon P, Wild F (1989) Expression of measles virus nucleoprotein in Escherichia coli: use of deletion mutants to locate the antigenic sites. J gen virol 70(Pt 2):435–441

Caignard G, Bourai M, Jacob Y, Tangy F, Vidalain PO (2009) Inhibition of IFN-alpha/beta signaling by two discrete peptides within measles virus V protein that specifically bind STAT1 and STAT2. Virology 383(1):112–120

Choi KS, Nah JJ, Ko YJ, Kang SY, Joo YS (2003) Localization of antigenic sites at the amino-terminus of rinderpest virus N protein using deleted N mutants and monoclonal antibody. J Vet Sci 4(2):167–173

Choi KS, Nah JJ, Ko YJ, Kang SY, Yoon KJ, Jo NI (2005) Antigenic and immunogenic investigation of B-cell epitopes in the nucleocapsid protein of peste des petits ruminants virus. Clin Diagn Lab Immunol 12(1):114–121

Dechamma HJ, Dighe V, Kumar CA, Singh RP, Jagadish M, Kumar S (2006) Identification of T-helper and linear B epitope in the hypervariable region of nucleocapsid protein of PPRV and its use in the development of specific antibodies to detect viral antigen. Vet Microbiol 118(3–4):201–211

Griffin DE, Pan C-H (2009) Measles: Old Vaccines. New Vaccines Curr top microbiol immunol 330:191–212

Heaney J, Barrett T, Cosby SL (2002) Inhibition of in vitro leukocyte proliferation by morbilliviruses. J Virol 76(7):3579–3584

Hussey GD, Clements CJ (1996) Clinical problems in measles case management. Ann Trop Paediatr 16(4):307–317

Jagtap SP, Rajak KK, Garg UK, Sen A, Bhanuprakash V, Sudhakar SB, Balamurugan V, Patel A, Ahuja A, Singh RK, Vanamayya PR (2012) Effect of immunosuppression on pathogenesis of peste des petits ruminants (PPR) virus infection in goats. Microb Pathog. doi:10.1016/j.micpath.2012.01.003

Jones L, Giavedoni L, Saliki JT, Brown C, Mebus C, Yilma T (1993) Protection of goats against peste des petits ruminants with a vaccinia virus double recombinant expressing the F and H genes of rinderpest virus. Vaccine 11(9):961–964

Karp CL, Wysocka M, Wahl LM, Ahearn JM, Cuomo PJ, Sherry B, Trinchieri G, Griffin DE (1996) Mechanism of suppression of cell-mediated immunity by measles virus. Science 273(5272):228–231

Koyama S, Ishii KJ, Coban C, Akira S (2008) Innate immune response to viral infection. Cytokine 43(3):336–341

Kumar A, Singh SV, Rana R, Vaid RK, Misri J, VS V (2001) PPR outbreak in goats: Epidemiological and therapeutic studies. Indian J Anim Sci 71:815–818

Laine D, Bourhis JM, Longhi S, Flacher M, Cassard L, Canard B, Sautes-Fridman C, Rabourdin-Combe C, Valentin H (2005) Measles virus nucleoprotein induces cell-proliferation arrest and apoptosis through NTAIL-NR and NCORE-FcgammaRIIB1 interactions, respectively. J gen virol 86(Pt 6):1771–1784

Laine D, Trescol-Biemont MC, Longhi S, Libeau G, Marie JC, Vidalain PO, Azocar O, Diallo A, Canard B, Rabourdin-Combe C, Valentin H (2003) Measles virus (MV) nucleoprotein binds to a novel cell surface receptor distinct from FcgammaRII via its C-terminal domain: role in MV-induced immunosuppression. J Virol 77(21):11332–11346

Libeau G, Diallo A, Calvez D, Lefevre PC (1992) A competitive ELISA using anti-N monoclonal antibodies for specific detection of rinderpest antibodies in cattle and small ruminants. Vet microbiol 31(2–3):147–160

Meng X, Dou Y, Zhai J, Zhang H, Yan F, Shi X, Luo X, Li H, Cai X (2011) Tissue distribution and expression of signaling lymphocyte activation molecule receptor to peste des petits ruminant virus in goats detected by real-time PCR. J Mol Histol 42(5):467–472

Mitra-Kaushik S, Nayak R, Shaila MS (2001) Identification of a cytotoxic T-cell epitope on the recombinant nucleocapsid proteins of Rinderpest and Peste des petits ruminants viruses presented as assembled nucleocapsids. Virology 279(1):210–220

Mondal B, Sreenivasa BP, Dhar P, Singh RP, Bandyopadhyay SK (2001) Apoptosis induced by peste des petits ruminants virus in goat peripheral blood mononuclear cells. Virus Res 73(2):113–119

Moussallem TM, Guedes F, Fernandes ER, Pagliari C, Lancellotti CL, de Andrade HF, Duarte MI Jr (2007) Lung involvement in childhood measles: severe immune dysfunction revealed by quantitative immunohistochemistry. Hum Pathol 38(8):1239–1247

Nanda SK, Baron MD (2006) Rinderpest virus blocks type I and type II interferon action: role of structural and nonstructural proteins. J Virol 80(15):7555–7568

Niewiesk S, Schneider-Schaulies J, Ohnimus H, Jassoy C, Schneider-Schaulies S, Diamond L, Logan JS, ter Meulen V (1997) CD46 expression does not overcome the intracellular block of measles virus replication in transgenic rats. J Virol 71(10):7969–7973

Opal SM, DePalo VA (2000) Anti-inflammatory cytokines. Chest 117(4):1162–1172

Raghavendra L, Setty DRL, Raghavan R (1997) Haematological changes in sheep and goats experimentally infected with Vero cell adapted peste des petits ruminants (PPR) virus. Indian J Anim Sci 12:77–78

Rajak KK, Sreenivasa BP, Hosamani M, Singh RP, Singh SK, Singh RK, Bandyopadhyay SK
(2005) Experimental studies on immunosuppressive effects of peste des petits ruminants
(PPR) virus in goats. Comp Immunol Microbiol Infect Dis 28(4):287–296

Ravanel K, Castelle C, Defrance T, Wild TF, Charron D, Lotteau V, Rabourdin-Combe C (1997)
Measles virus nucleocapsid protein binds to FcgammaRII and inhibits human B cell antibody
production. J Exp Med 186(2):269–278

Renukaradhya GJ, Sinnathamby G, Seth S, Rajasekhar M, Shaila MS (2002) Mapping of B-cell
epitopic sites and delineation of functional domains on the hemagglutinin-neuraminidase
protein of peste des petits ruminants virus. Virus Res 90(1–2):171–185

Romero CH, Barrett T, Chamberlain RW, Kitching RP, Fleming M, Black DN (1994)
Recombinant capripoxvirus expressing the hemagglutinin protein gene of rinderpest virus:
protection of cattle against rinderpest and lumpy skin disease viruses. Virology
204(1):425–429

Roulston A, Marcellus RC, Branton PE (1999) Viruses and apoptosis. Annu Rev Microbiol
53:577–628

Sahinduran S, Albay MK, Sezer K, Ozmen O, Mamak N, Haligur M, Karakurum C, Yildiz R
(2012) Coagulation profile, haematological and biochemical changes in kids naturally
infected with peste des petits ruminants. Trop Anim Health Prod 44(3):453–457

Schlender J, Schnorr JJ, Spielhoffer P, Cathomen T, Cattaneo R, Billeter MA, ter Meulen V,
Schneider-Schaulies S (1996) Interaction of measles virus glycoproteins with the surface of
uninfected peripheral blood lymphocytes induces immunosuppression in vitro. Proc Nat Acad
Sci USA 93(23):13194–13199

Schnorr JJ, Seufert M, Schlender J, Borst J, Johnston IC, ter Meulen V, Schneider-Schaulies S
(1997) Cell cycle arrest rather than apoptosis is associated with measles virus contact-
mediated immunosuppression in vitro. J gen virol 78(Pt 12):3217–3226

Scott GR (1981) Rinderpest and peste des petits ruminants. In: Virus diseases of food animals, vol
2. Academic Press, London

Sinnathamby G, Renukaradhya GJ, Rajasekhar M, Nayak R, Shaila MS (2001) Immune
responses in goats to recombinant hemagglutinin-neuraminidase glycoprotein of Peste des
petits ruminants virus: identification of a T cell determinant. Vaccine 19(32):4816–4823

Sinnathamby G, Seth S, Nayak R, Shaila MS (2004) Cytotoxic T cell epitope in cattle from the
attachment glycoproteins of rinderpest and peste des petits ruminants viruses. Viral Immunol
17(3):401–410

Svitek N, von Messling V (2007) Early cytokine mRNA expression profiles predict Morbillivirus
disease outcome in ferrets. Virology 362(2):404–410

van Riel D, Leijten LM, van der Eerden M, Hoogsteden HC, Boven LA, Lambrecht BN,
Osterhaus AD, Kuiken T (2011) Highly pathogenic avian influenza virus H5N1 infects
alveolar macrophages without virus production or excessive TNF-alpha induction. PLoS
Pathog 7(6):e1002099

Vaux DL, Strasser A (1996) The molecular biology of apoptosis. Proc Nat Acad Sci USA
93:2239–2244

von Andrian UH, Mackay CR (2000) T-cell function and migration. Two sides of the same coin.
N Engl J Med 343(14):1020–1034

White E (1996) Life, death, and the pursuit of apoptosis. Genes Dev 10:1–15

Yarim GF, Nısbet C, Yazıcı Z, Gumusova SO (2006) Elevated serum total sialic acid
concentrations in sheep with peste des petits ruminants. Medycyna Weterynaryjna
62(12):1375–1377

# Chapter 5
# Epidemiology and Distribution of Peste des Petits Ruminants

**Abstract** Phylogenetically, based on the fusion (F) and nucleocapsid (N) genes, peste des petits ruminants virus (PPRV) can be classified into four distinct lineages. PPRV belonging to lineages I and II are exclusively isolated from the countries in West Africa where PPRV originated. Lineage III is restricted to the Middle East and East Africa. Lineage IV is considered to be a new lineage comprising newly emerging viruses, and is currently most prevalent in Asian countries and becoming overwhelmed lineage in Africa. It is not clear whether the apparent geographic spread of the disease in the last 50 years is real, or whether it reflects increased awareness, wider availability of diagnostic tools or even a change in the nature of the virus. It seems most likely that a combination of factors contribute to the current epidemiologic pattern of the disease. Confusion of PPR with pneumonic pasteurellosis and other pneumonic diseases of small ruminants has delayed its recognition in some countries. Nevertheless, the proper understanding of the lineage distribution in a specified region is essential when choosing the appropriate homologous prototype stain for vaccine production to ensure efficient immunization. Continued application of heterologous vaccine candidates hitherto not prevalent may lead to generation of novel lineages, or allow the existing population to evade protection, especially for RNA viruses. Therefore, identification of the lineage is a prerequisite for fruitful diagnosis, epidemiology and control. The current known disease distribution of PPRV is discussed comprehensively in this chapter.

**Keywords** Epidemiology · Distribution · Phylogenetic analysis · F gene · N gene

M. Munir et al., *Molecular Biology and Pathogenesis of Peste des Petits Ruminants Virus*, SpringerBriefs in Animal Sciences, DOI: 10.1007/978-3-642-31451-3_5, © The Author(s) 2013

## 5.1 Introduction

Due to the increased importance of infectious diseases of animals during the past 10 years, avian influenza, FMD, PPR, bluetongue, rabies and others have been revealed. PPR takes special place among those diseases that affect goats and sheep. After the first report of PPR in the Ivory Coast, West Africa in 1942 (Gargadennec and Lalanne 1942) different names such as "kata", "pseudo rinderpest", "syndrome of stomatitis-pneumoenteritis" and "ovine rinderpest" have been used to describe the disease. It was not clear until the French name "*peste des petits ruminants*" was used because of its clinical, pathologic and immunological similarities with rinderpest. In the following four decades, until 1979, PPR was confirmed in most countries in West Africa, such as Nigeria, Senegal, Togo, and Benin, while by 1982 the disease had gone as far as Sudan, an eastern African country. Epizootiological investigations have indicated that PPRV has broadened its territory toward south of Africa since its first report. Unrestricted movement of PPR susceptible animals within West African countries could possibly be the reason for this directional spread.

Despite the antigenic and immunological similarity to RP, PPR remained an unrecognized and undiagnosed disease for long time after its first recognition. India was the first country in Asia where PPR was diagnosed, in 1987 (Shaila et al. 1989), followed by Pakistan, where it was first been reported in 1994 (Amjad et al. 1996). From 1993 to 1995, PPRV spread lavishly among the countries of the Arabian Peninsula, South Asia and the Middle East, where it has since remained an endemic disease. At present, PPR is serologically confirmed in most of the countries of the African continent, Central, Middle and South Asia, the Middle East and the Arabian Peninsula (Fig. 5.1). Recently, Kaukarbayevich (2009) revealed that the epizootic situation of PPR disease was worst from 1986 to 1999, when 50–70 outbreaks were reported out of every 10 millions head of small ruminants, while the situation became a little more favourable when this severity was reduced to 10–30 outbreaks in recent years (Kaukarbayevich 2009).

Phylogenetically, PPRV can be classified into four lineages (Shaila et al. 1996; Dhar et al. 2002). PPRV belonging to lineage I and II are exclusively isolated from the countries of PPRV origin in West-Africa. Lineage III is restricted to Arabia and East Africa, although some of the viruses that belong to lineage III have also been isolated from southern India. Lineage IV is considered a new lineage comprising newly emerging viruses. Surprisingly, this lineage is very close to lineage I, which is a typical African lineage. Through an unknown source, lineage IV succeeded to invade Asia and the Middle East.

There has been a substantial expansion of PPR worldwide over the last three decades, and recently it has been diagnosed in already known enzootic areas (Arzt et al. 2010). In Asia, PPRV has recently been reported for the first time in China, Nepal and Tajikistan, while in Africa PPRV has now expanded from South of the Equator to Gabon (1996), Kenya (2006), the Congo (2006) and Uganda (2007), and also to the north of the Sahara to Morocco (2007) (ProMED 2008), which indicates its continuous threat around the globe (Fig. 5.1). Sequence

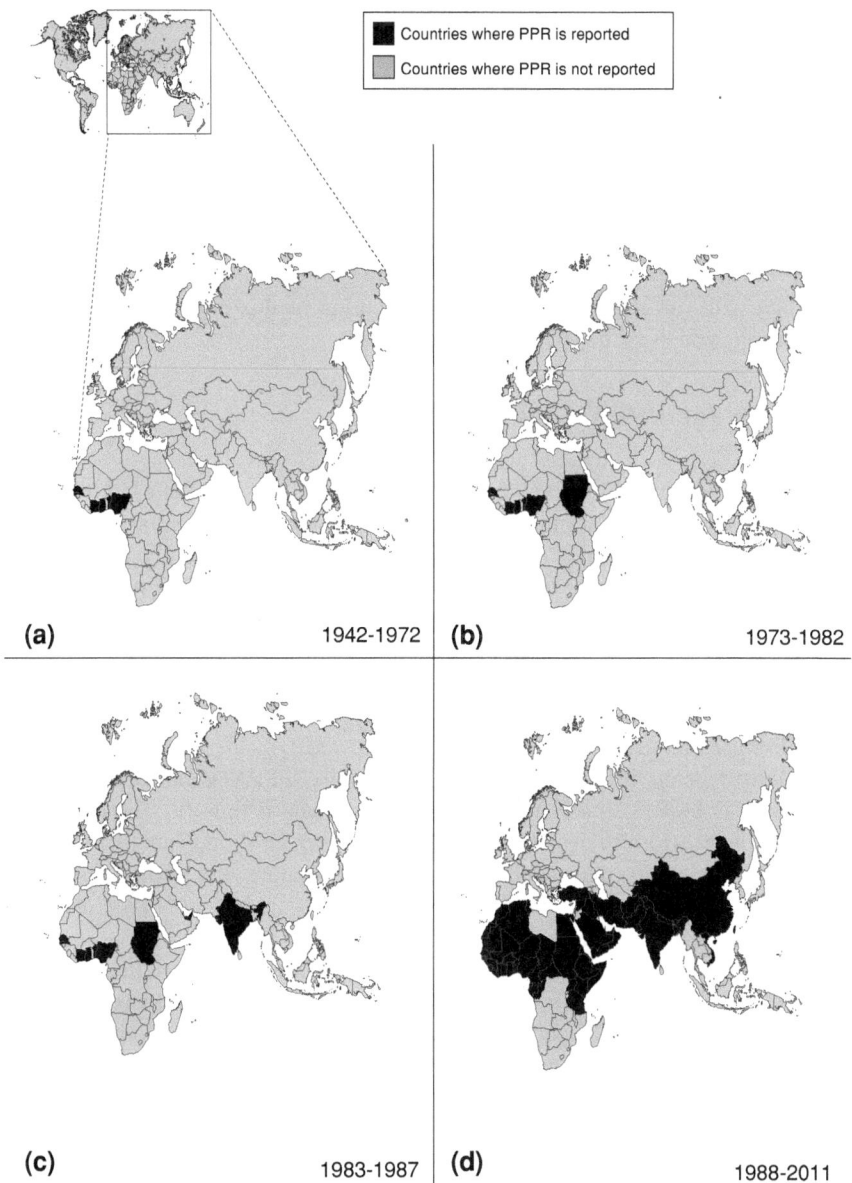

**Fig. 5.1** Overview of the geographic distribution of PPR around the globe. This map was created based on the reports of PPR to OIE, based on both serologic and genomic detection. Note the disease spread from country of origin (**a**) in all directions (**b, c**), and reached to most of the countries of Asia and Africa (**d**)

analysis of the PPRV N gene of an isolate from Morocco indicates that it is closely related to Saudi Arabian and Iranian strains. This suggests that the spread of the virus to Morocco was due to the enormous import of small ruminants from the Middle East. Taking intercontinental spread of the virus into account, it is likely that new African lineages may be occurring in Morocco (ProMED 2008). Despite the fact that PPR has been restricted to Africa, Asia and the Middle East, it has expanded in the past 10 years. Such findings have highlighted how important it is to understand the way that these viruses are restricted in the range of a region in which they cause disease, an understanding that will become increasingly important with the success of possible eradication of this disease on a local and global level, which is the focus of this chapter.

## 5.2  Bases for PPRV Classification

On the basis of phylogenetic analysis of morbilliviruses, it is believed that when cattle domesticated they contained a morbillivirus, which appeared as a progenitor of modern rinderpest virus (RPV). Furthermore, RPV ultimately evolved into measles virus (MV), which latter adapted to humans. The conversion of RPV to canine distemper virus (CDV) is believed to occur when carnivores ate the ruminants, thus acquired the morbillivirus infection, which then evolved into CDV (Barrett 1999). MV and RPV are described as closely related, and CDV and phocine distemper virus are the most distantly related to MV and RPV among morbilliviruses (Barrett 1999). PPRV exhibited the typical characteristics of the Morbillivirus genus in the Paramyxoviridae family. PPRV is not only a distinct virus but may be less closely related to RPV than MV is to RPV. The other three members of the Morbillivirus genus (MV, CDV, and RPV) indicate that strains of varying pathogenicity may occur naturally. Furthermore, the strains distinguish themselves from virulent strains by including a faster migrating N protein (Diallo et al. 1987) or by their MAb reactivity range (Libeau et al. 1992).

The use of phylogenetic analysis to clarify epidemiologic relationships, and possibly identify changes in pathogenicity or host preference, has become a valuable tool. Based on the partial sequence analysis of the fusion protein (F) gene, PPRV can genetically be divided into four distinct lineages named I, II, III and IV (Fig. 5.2a) (Dhar et al. 2002). However, there is only a single serotype reported. The F gene based classification system was implemented from the beginning of PPRV identification, and has broadened our understating of the molecular epidemiology of the disease along with movement and distribution of the virus. Continuous circulation of the virus in endemic countries, and reports of PPRV in previously disease free countries, demand that we delve into the molecular details of the field virus.

In this regard, Kerur et al. (2008) made a parallel comparison of both the traditional F gene and targeting the N gene for determining the molecular epidemiologic pattern on the same virus samples. They revealed the partial F gene sequence based classification of PPRV into lineages placed the studied isolates

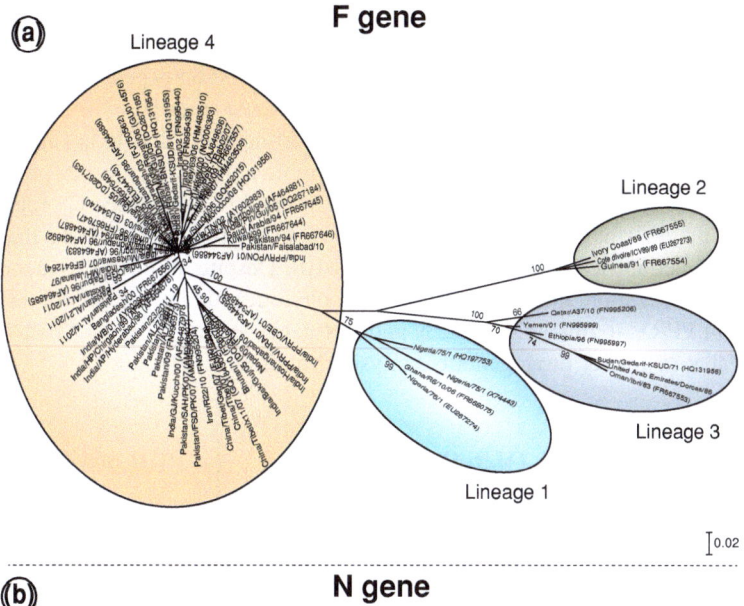

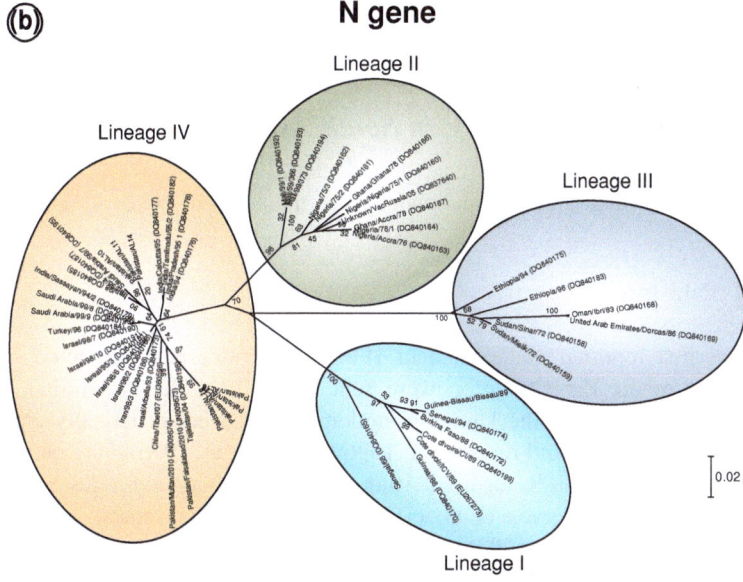

**Fig. 5.2** Majority rule consensus tree of PPR viruses based on the F gene (**a**) and N gene (**b**). The tree was constructed using the neighbour-joining method using the Kimura-two-parameter model in MEGA4 version 4. Numbers indicate the bootstrap values (1000 replicates), and only values above 50 % are shown in the figure. Horizontal distances are proportional to sequence distances. The figure shows the clear division of PPRV into four lineages

into lineage IV, whereas classification of PPRV into different lineages based on N gene sequence (lineage I, II, III, IV) appeared to group the viruses in a better way, thus, giving a better epidemiologic picture about PPRV (Kerur et al. 2008)

(Fig. 5.2b). However, all the PPRV strains remained in the same group regardless of the gene used, except that PPRV strains belonging to lineage I (i.e. Nig/75) based on the F gene appeared instead as lineage II based on N gene tree. Currently, Balamurugan et al. (2010) compared the phylogenetic trees based on the N, F, M and HN genes of Indian origin PPRV. They concluded that it is important to monitor the circulation of the PPRV in susceptible animals by the HN gene-based sequence comparisons in addition to the F gene- and N-gene based approaches, to identify the distribution and spread of the virus in the regular outbreaks that occur in endemic countries (Balamurugan et al. 2010). Despite all of this evidence, it is desirable to use more than one viral gene for phylogenetic interpretation, due to the ability of PPRV to mutate. The F, N and HN genes appear to be the most suitable candidates so far for phylogenetic analysis (Munir et al. 2012b).

## 5.3 Prevalence and Distribution of PPRV in PPR Endemic Countries

PPR has been described in most of the countries of the Asian and African continents and the epidemiology is discussed in detail in each of these countries.

### 5.3.1 Distribution of PPRV in South Asia

PPRV is widespread and remains endemic in most of the South Asian countries (Fig. 5.3).

#### 5.3.1.1 Pakistan

PPR was first reported in Pakistan in 1991, when rinderpest suspected samples from goats in the Punjab region were sent to the Institute for Animal Health, Pirbright Laboratory, in the United Kingdom, which were genetically characterized as PPRV in 1994 and reported in 1996 (Amjad et al. 1996). However, before this confirmation of PPR, again a rinderpest-like disease was described from Punjab province by Pervez et al. (1993) in the Pakistan Journal of Livestock Research (Pervez et al. 1993), entirely based on clinical outcome. With the confirmation of PPRV it is likely that the suspected outbreaks, if not all, were due to PPRV infection and might be prevalent much earlier. Based on nucleotide analysis of the F gene, it was revealed that the Pakistani PPRV show relationship to Iran/94, Bangladesh/94 and India/94 in descending order, with least relationship to Nigeria75/1, which was the prevailing strain of PPRV at that time.

In the following years, with the introduction of serodiagnostic tests, rumours of rinderpest eradication and awareness of PPRV, the disease was reported from several districts of Punjab and from governmental livestock farms (Khan et al. 2008; Abubakar et al. 2008; Munir et al. 2009). However, a significant

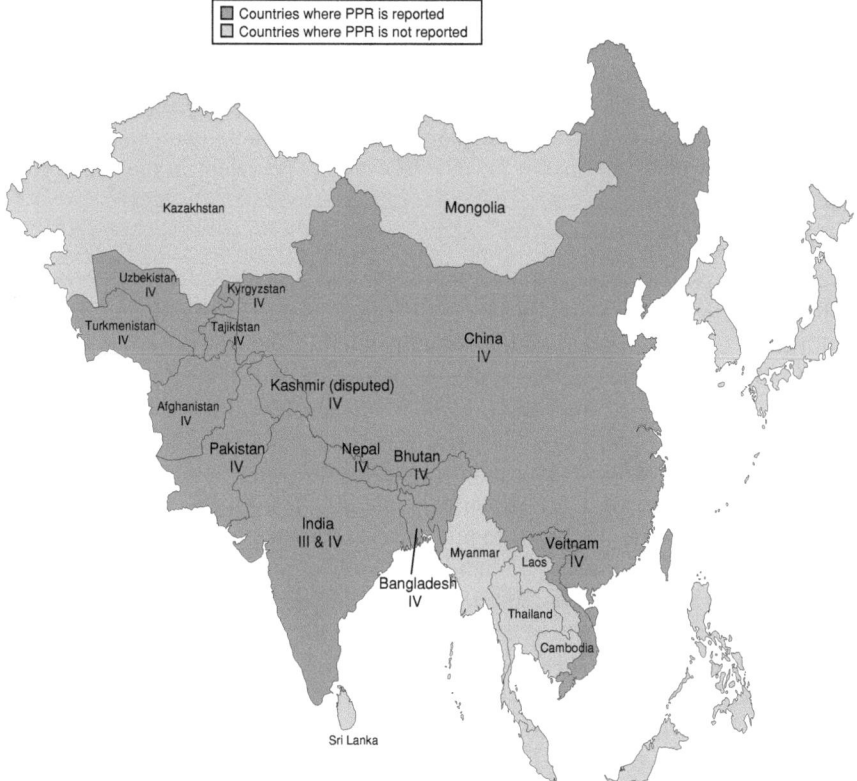

**Fig. 5.3**  Epidemiology and distribution of PPRV in South Asia

contribution was made by a FAO funded study in which samples were collected from the whole country and were serologically (cELISA) analysed (Zahur et al. 2008). In this study, 1463 samples from sheep and goats were collected from 17 selected districts from all four provinces [Punjab, Khyber Pakhtunkhwa (former North Western Frontier Province), Sindh, Baluchistan], Islamabad Capital Territory, Azad Jammu and Kashmir (AJK), Northern Areas, and local and Afghani nomads. The disease was sero-diagnosed using cELISA in all of the selected areas, which range from 7.1 to 100 % with an overall positivity of 74.9 %. Thus, this study has clearly provided evidence that the disease is prevalent throughout country in both sheep and goats.

The diagnosis of PPRV from Pakistan, in most of the above mentioned and other studies, was based on either clinical assessment or serodiagnosis. Only a few sequences from the F gene of PPRV are available in GenBank, which were not sufficient for establishing epidemiologic links between current outbreaks. Moreover, due to the shift in interest from the F gene to the N gene for phylogenetic analysis, we characterized PPRV from sheep and goats from Pakistan based on the N gene for the first time, and further expanded the availability of sequences for the

F genes (Munir et al. 2012b). In this report, it was possible to conclude that based on the F gene the Pakistani isolates clustered with the Kuwaiti and Saudi Arabian PPRV isolates whereas based on the N gene the Pakistani isolates appeared to be closely related to the Chinese, Tajikistani and Iranian isolates. Although intensive research is required to characterize the isolates from whole country, the current status indicates that it is lineage IV of PPRV that is prevalent in the country.

### 5.3.1.2 India

PPRV in India was first reported in 1987 from Tamil Nadu state, and remained confined to the same region until 1994 (Shaila et al. 1989). Following a solitary report of PPRV in buffalo in the same state (Govindarajan et al. 1997), the disease spread to other parts of the country, which was also the time when disease was reported in other neighbouring countries. Current reports have demonstrated that the disease in endemic in the country, and is now evident in the Thar desert (Rajasthan, a northern state) (Kataria et al. 2007), Kolkata, an eastern state (Saha et al. 2005), the Parbhani region (Karnataka state) (Chavran et al. 2009), Maharastra in the south-west of India (Santhosh et al. 2009), and the southern peninsula (Raghavendra et al. 2008).

There are several reports of systemic analysis of PPR and seroprevalence in small ruminants in India (Raghavendra et al. 2008; Singh et al. 2004). However, a current study provides an intensive overview of the disease in large ruminants (cattle and buffalo) in southern peninsular India. Analysis of 2159 samples from cattle and buffalo estimated a seroprevalence of 4.6 % which highlights the susceptibility of large ruminants for PPRV (Balamurugan et al. 2012).

Except a single report of lineage III in buffalo, all of the characterized Indian PPR viruses belong to lineage IV with non-significant variations between isolates (Dhar et al. 2002). It is speculated that this single report of lineage III was dried out and replaced with lineage IV (Banyard et al. 2010). If this is the case then it is extremely important to monitor those factors that are responsible for selection of a specific lineage of PPRV and mark the geographic restriction of the virus. This hypothesis is less likely the case in the PPRV scenario, since more than one lineage has been reported from the same region, such as lineage III and IV from Sudan and Qatar (Kwiatek et al. 2011).

### 5.3.1.3 China

The disease has first been reported in China in 2007 when sheep and goats were screened from the Tibet region (Shannan, Shigatse, Nagqu, Nyingchi, and Ngari) (Wang et al. 2009). However, it was speculated that the disease might prevail in Tibet due to a lack of awareness of the clinical outcome of the disease. Molecular characterisation of positive PPRV samples from Ngari indicated that all of the isolates belong to lineage IV, which was believed to be restricted to Southeast Asia. The overall topology of the tree indicated that these isolates closely related to Indian and Tajikistani isolates. Due to uncontrolled animal movements and unofficial trade between Tibet and bordering nations, it is possible that the disease might spread from neighbouring nations such as India and Nepal to Tibet (Wang et al. 2009). Recently, characterisation of PPRV in free-living bharals (*Pseudois nayaur*) indicated the circulation of PPRV belonging to lineage IV that are identical to previously characterized PPRV isolates from Tibet, which highlights the importance of wildlife in the epizootiology of the disease (Bao et al. 2011). With these reports, no further spread of the disease has been reported.

### 5.3.1.4 Bangladesh, Nepal and Sri Lanka

In Bangladesh, the disease was reported in the same period when it was recognized in Pakistan, in 1993, in black Bengal goats in the Mymensingh area (Islam et al. 2001). Recently, the same group identified the disease again in black Bengal goats by RT-PCR and made efforts to characterize the virus pathologically (Rahman et al. 2011). Genetic characterisation of Bangladeshi PPRV isolates demonstrated that these belong to lineage IV and are closely related to the Indian isolates. The disease was described in Nepal in 1995 and, interestingly, isolates from Nepal, Bangladesh and India made a distinct cluster that is different from the cluster constituted by the isolates from Pakistan, Saudi Arabia, Kuwait and Iran (Dhar et al. 2002), which probably reflects the close trade between these groups of countries. However, all PPRV isolates from these countries belong to same lineage, lineage IV. No official report of PPRV is available from Sri Lanka, which may be due to its unique geographic location with no land connection with neighbouring countries.

### 5.3.1.5 Afghanistan

Despite disease reports in most of the bordering countries of Afghanistan, it remained a matter of internal stability to screen the heavily populated areas with small ruminants in the country for the presence of PPRV. It is believed that the disease appeared in Afghanistan at the same when it was recorded in Pakistan. In this regard, during 1995–96 serum samples collected from Khost province for the rinderpest appeared positive for PPRV. However, officially it was not until 2003 when investigators from the Ministry of Agriculture and the FAO livestock programme in Kabul collected samples from sheep and goats from the Northern provinces of Afghanistan. A high seropositive (42/46) sample for PPRV was detected, which was supported by the clinical picture of the animals identical to PPR (Martin and Larfaoui 2003). However, it was not possible to rule out that this seropositivity was due to vaccination, because of unavailability of clinical samples for genome detection of PPRV. Later, it was estimated that there were 7741 cases of PPRV in sheep and goats in only 15 provinces of Afghanistan. Furthermore, competitive ELISA has been applied to ascertain the seroprevalence of PPRV in 60 villages from 17 provinces. A high seropositive rate ($n = 790$) was observed from the collected samples ($n = 4048$), both in sheep and goats (Dr. Nawroz, unpublished data). Currently, we have collected samples from nomads' sheep and goats entering into Pakistan for sale for "Eid-ul-Adha" (a Muslim religious tradition to sacrifice animals for the sake of God). Genetic characterisation indicated that PPRV from Afghani nomads belong to lineage IV, with substantial differences from the isolates characterized from Pakistan at the same time (M. Munir, unpublished data).

### 5.3.1.6 Kazakhstan

PPR was first described in Kazakhstan when a large number of samples was analysed at the Institute of Animal Health, UK, which were collected from cattle, sheep and goats during Post-Soviet Transitions from 1997 to 1998. A low number of cattle (6/279), sheep (3/542) and goat (1/137) were found seropositive with cELISA (Lundervold et al. 2004). The disease has been monitored and reported later; however, the information regarding the genetic nature is till lacking to ascertain which lineage of PPRV is circulating in Kazakhstan, if not lineage IV as expected by virtue of its geography.

### 5.3.1.7 Tajikistan

After a transit period of misdiagnosis of PPR as pasteurellosis (a clinically similar disease caused by *Pasteurella* spp.), PPR was confirmed for the first time in three districts (Gharm, Farkhror and Tavildara) of Tajikistan. First, the confirmation was made based on serology by application of cELISA, and then genetically characterized by sequencing and phylogenetic analysis of the N gene of PPRV. The tree topology indicated that PPRV isolates from Tajikistan clustered in lineage IV, as expected, and were closely related to the Iranian and Saudi Arabian PPRV isolates (Kwiatek et al. 2007) sequenced at that time. However, characterisation of PPRV isolates from Pakistan indicated that Tajikistani PPRV isolates are the closest relative to the Pakistani PPRV isolates (Munir et al. 2012b). Historically, the outbreak appeared in Gharm district of Tajikistan in goats that were imported from the eastern districts of Tajikistan bordering Afghanistan and China. The disease was officially not reported in China and Afghanistan at that time but it was speculated that PPRV existed in the region. Notably, a disease report and characterisation of PPRV in China later in 2007 revealed that Chinese PPRV isolates clustered closest to PPRV isolates from Tajikistan (Munir et al. 2012b; Wang et al. 2009). It is therefore logical to postulate that the genetic nature of PPRV in China, Afghanistan, Iran, Tajikistan and other adjacent and bordering countries is slightly variable and may have a common origin, possibly from Saudi Arabia being part of same cluster.

### 5.3.1.8 Vietnam and Bhutan

Owing to bordering China at the north, serologic analysis indicated that the disease is prevalent in Vietnam, probably at the same time when it was reported from China in 2007 (Maillard et al. 2008). In this analysis, which was conducted on 283 goats, 63 cattle and 22 buffalo, a relatively low level of seropositive cases (3, 1, 1 respectively) was observed. Interestingly, no clinical disease was noticed before the sampling and one year after the analysis; however, these animals remained positive serologically when tested again with cELISA after 1 year. It was concluded that the co-existence of domestic and wildlife in the same ecosystem preserved the biodiversity and established equilibrium of co-adaptation between pathogens and their hosts. This leads to generation of genetic resistance to the pathogens. It remains to be explored whether the genetic nature of these circulating viruses might be adapted enough to cause immunogenicity but not pathogenicity, and can act as a model for the genesis of viruses with reduced virulence but retaining immunogenicity. Recently, material submitted to the Regional Reference Laboratories has confirmed the presence PPRV in Bhutan, and that these isolates belong to lineage IV of PPRV (Banyard et al. 2010).

## 5.3.2 Distribution of PPRV in the Middle East

Most of the countries of the Middle East are reported for the presence of PPRV (Fig. 5.4). The situation of PPRV in each of the countries is discussed in detail below.

### 5.3.2.1 Iraq

Although the disease has been observed for several years, and when it was known to be present in other neighbouring countries, it was in 1998 when PPR in Iraq was officially reported to OIE and FAO (FAO 2003). In 1999 during the onset of disease outbreaks, Dr. Samir Hafez and Dr. Adama Diallo have started mass vaccination and made efforts to strengthen the diagnostic capacities of the practicing veterinarians in 12 governorates. However, the official report indicates that the disease occurred in 2000 where it caused high mortality in sheep (Barhoom et al. 2000). Recently, a devastating outbreak of PPRV in wild goats (*Capra aegagrus*) caused 762 deaths (354 males, 408 females) in just 7 months. Genetic characterisation of the isolates at the Friedrich-Loeffler-Institute, Germany, indicated that these belong to lineage IV of PPRV and clustered close to the Turkish PPRV isolates (Hoffmann et al. 2012). It is interesting to observe that the disease was restricted to only wild animals and did not cause disease to domesticated animals, probably due to vaccination, which prevented the spill over of the virus.

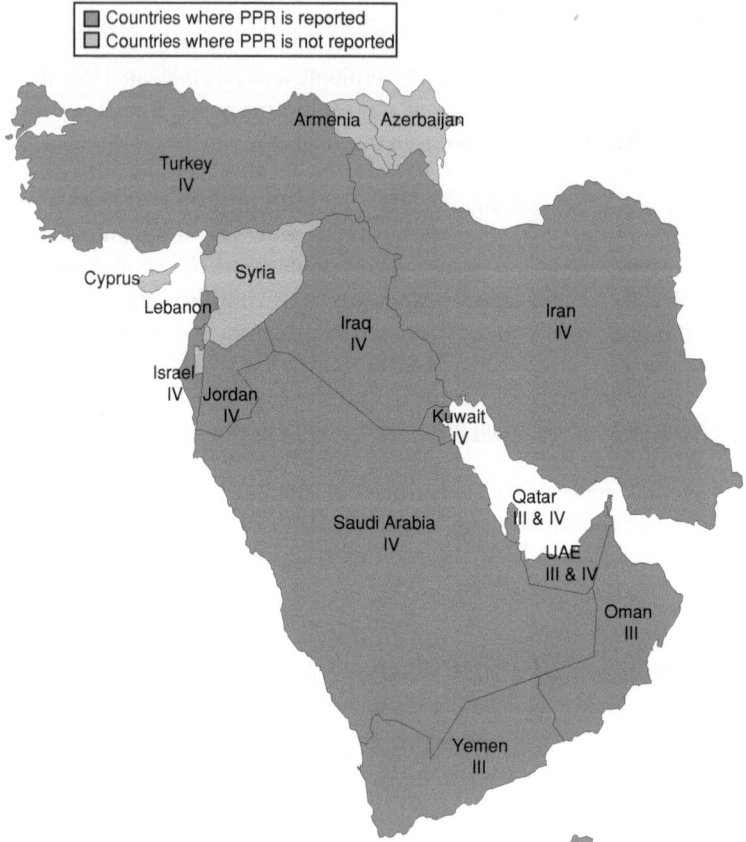

**Fig. 5.4** Epidemiology and distribution of PPRV in Middle East

### 5.3.2.2 Iran

Based on clinical, pathologic, and serologic documentation, it is evident that PPR in Iran dates back to 1995, when a specific disease was observed in Ilam province of Iran (Radostits et al. 2000). After this, the disease was reported from most of the provinces ($n = 28$) over a period of 10 years (1995–2004). During this period, there were around 1433 flocks affected, in which the disease occurrence was highest in Gom province ($n = 283$ flocks affected), whereas the lowest was in Semnan province ($n = 3$ flocks affected) (Bazarghani et al. 2006). Latter studies revealed that the Iranian PPRV isolates belong to lineage IV and are closely related to the Pakistan, Saudi Arabian, Tajikistan, and Chinese isolates of PRPV (Esmaelizad et al. 2011; Kwiatek et al. 2007; Munir et al. 2012b).

### 5.3.2.3 Israel

Israel is a country where PPRV was diagnosis in the 1990s (Perl et al. 1994); and molecular typing indicated that the Israeli strains of PPRV cluster close to the Turkish isolates (Israel/95) or branched distantly enough to be considered a separate subcluster (Israel/98) (Banyard et al. 2010; Munir et al. 2012a, b, c).

### 5.3.2.4 Saudi Arabia

The clinical and serologic evidence of PPR's presence in Saudi Arabia date back to the 1980s, when the disease was described in sheep and wild ruminants (deer and gazelles) in 1980 and 1987, respectively (Asmar et al. 1980; Hafez et al. 1987). However, the genetic nature of the virus was not determined nor was the virus isolation successful until 1990, when the first virus isolation of PPRV became possible in a goat outbreak in Al-Ahsa oasis in the East of the country (Abu Elzein et al. 1990). The disease was observed throughout the 1980s and later. In early 2002, the disease emerged in both sheep and goats in Al-Hasa province in the eastern region of Saudi Arabia, which caused 100 % mortality. Several diagnostic tests, including an agar gel immunodiffusion test, virus neutralization test and fluorescent antibody test confirmed that the disease was caused by PPRV (Housawi et al. 2004).

Recently, the seroprevalence was determined in ten governorates in the central region of Saudi Arabia based on samples collected from September 2005 to March 2006. It was revealed that there was a high prevalence of PPRV in sheep (363/992 = 36.59 %) and in goats (530/962 = 55.09 %) (Al-Dubaib 2008). Examined cattle and camels that were grazing with these seropositive small ruminants were found to be negative for antibodies against PPRV when determined by cELISA, eliminating the role of camels in the disease transmission, which was earlier believed to happen (Roger et al. 2001b). The genetic characterisation of the isolates indicated that these cluster close to the Pakistani, Irani, and some of Chinese and Tajikistani isolates within lineage IV (Kwiatek et al. 2007).

### 5.3.2.5 Sultanate of Oman and Yemen

It was in the Sultanate of Oman where PPRV was, for the first time, reported outside Africa, in a countrywide survey carried out in 1978 (Hedger et al. 1980). A few years later, Taylor et al. (1990) determined the epidemiology of PPRV in sheep and goats in four regions (Batina coast, Oman interior, Sharqiyah and Salalah), and detected a seropositivity of 26.5 %, 32.8 %, 24.5 %, and 4.8 %, respectively. Additionally, they made a substantial contribution to the basic serology and differential diagnosis of PPRV from RPV, which was intermittently observed in the region at that time (Taylor et al. 1990). Based on PAGE mobility pattern, they observed that Omani PPRV isolates show a pattern that is distinct from African PPRV isolates, which was also seen in virus neutralization abilities of both type of viruses. These observations led them to speculate that African and Omani PPRV isolates probably evolved independently, due to long period of physical separation from same mother virus, rinderpest. Later, phylogenetic analysis of the N and F genes indicated that the Omani isolates belong to lineage III, which shows high identity to that of PPRV isolates characterized from the United Arab Emirates (UAE) (Kwiatek et al. 2011). In Yemen, a most southern region of Abranian peninsula, lineage III has been circulating since its first identification (Dhar et al. 2002). Interestingly, being close to Saudi Arabia (where lineage IV PPRV were isolated), no report describes the circulation of lineage IV in either Yemen or Oman (Banyard et al. 2010).

### 5.3.2.6 United Arab Emirates

The report of Furley et al. (1987) identified PPRV in a variety of zoo animal species, including gazelles (*Gazellinae*), ibex and sheep (*Caprinae*) and gemsbok (*Hippotraginae*) and Nilgai (*Tragelaphinae*) at the end of 1983 (Furley et al. 1987). Additionally, this report defined the host range of PPRV in the zoological collection, and made a substantial contribution to knowledge of the highly infectious nature of this virus. At the same time, a study attempted to describe the incidence of PPR-like diseases from 1987 to 1989 in the Al-Ain region, UAE, which contributed approximately one-third of the total national livestock population. It was found that there were at least 4, 15 and 22 outbreaks of PPR in 1987, 1988 and 1989, respectively, in which the disease remained most prominent in the month of July (Moustafa 1993). A survey, in which 294 sera were collected during 1999–2001 from eight captive and one free-ranging herds of Arabian Oryx (*Oryx leucoryx*), indicated no sign of PPRV in the UAE. Molecular characterisation revealed that the virus strains isolated in the 1980s belonged to lineage III of PPRV. However, a recent study conducted by Kinne et al. (2010) in a wide range of Arabian wildlife species indicated that PPRV characterized from these animals belong to lineage IV, and is closely related to a recently characterized strain from China, but is distant from lineage IV strains originating in gazelles isolated in 1999 and 2002 from Saudi Arabia. Moreover, it is not related to the lineage III strains isolated from a Dorcas gazelle in the UAE in 1986 (Kinne et al. 2010). Although the origin of this novel strain remains elusive, it was speculated that importation of infected domestic or wild small ruminants from Asia into the UAE or other countries of the Arabian Peninsula has to be considered as a possible source of infection (Kinne et al. 2010).

### 5.3.2.7 Qatar

Similar to the UAE, both lineages III and IV were characterized from Qatar in 2010. PPRV has recently been identified in the wild deer population, which explains the crucial role of wildlife in the epizootiology of the disease [C. Oura unpublished data described in (Banyard et al. 2010)]. The rate of incidence of

PPRV in Qatar is low compared to other neighbouring countries, probably due to its distinct geographic location.

### 5.3.2.8 Lebanon

Despite the fact of clinical evidence of PPRV prevalence in the Bekaa and South Lebanon districts (Hermel, Baalbeck, Tyre and Saida regions), the disease has only recently been serologically monitored. In a study conducted on 2205 goat and 1300 sheep blood sera collected from 20 districts, Hilan et al. (2006) reported a seroprevalence of 52.0 % and 61.5 % in goat and sheep individuals, respectively. Additionally, the cow showed a seroprevalence of 5.72 % (Hilan et al. 2006). It was also noticed that Bekaa and South Lebanon are the most effected areas among the four tested regions in Lebanon. Later, Attieh reported seroprevalence of up to 48.6 % in Lebanon (Attieh 2007).

### 5.3.2.9 Kuwait

PPRV reported from Kuwait (Kuwait/99) belong to lineage IV and are closely related to the Saudi Arabian PPRV strains isolated during 1994. A study conducted by Dhar et al. (2002) revealed an interesting relationship between PPRV strains of Pakistan, Saudi Arabia, Kuwait and Iran, where these form a cluster which is distinct from the cluster made by the PPRV isolates of India, Nepal and Bangladesh. Based on this, it was concluded that it is likely that the same virus that is circulating in these west Asian countries, and that they might have a common origin (Dhar et al. 2002).

### 5.3.2.10 Turkey

In Turkey, PPRV was first reported in September 1999 when goats in Elazig Province, eastern Anatolia, succumbed to the virus (EMPRES 2000). This was the first outbreak of PPR in Turkey ever reported to OIE; however, it is believed that the disease might have been present even before (Alcigir et al. 1996; Tatar 1998). Later, disease outbreaks appeared in seven villages located near the city of Bursa in the Marmara region of western Turkey between July 2002–September 2003. The clinical and molecular findings of this study demonstrated the presence of PPR in the Bursa province in western Anatolia, close to the European territory of Turkey (Yesilbag et al. 2005). Later, several studies reported the disease in other parts of the Turkey. In only 2005, around 78 separate outbreaks were recorded throughout Turkey, which dictates quarantine and vaccination to prevent further spread of the disease. In 2007, Kul et al. (2007) reported the disease of PPRV in Kirikkale Province, Central Anatolia, which provided evidence that the disease is present in central Turkey (Kul et al. 2007). The report of Albayrak and Alkan (2009) suggested that PPRV is prevalent in the Middle and Eastern Black sea

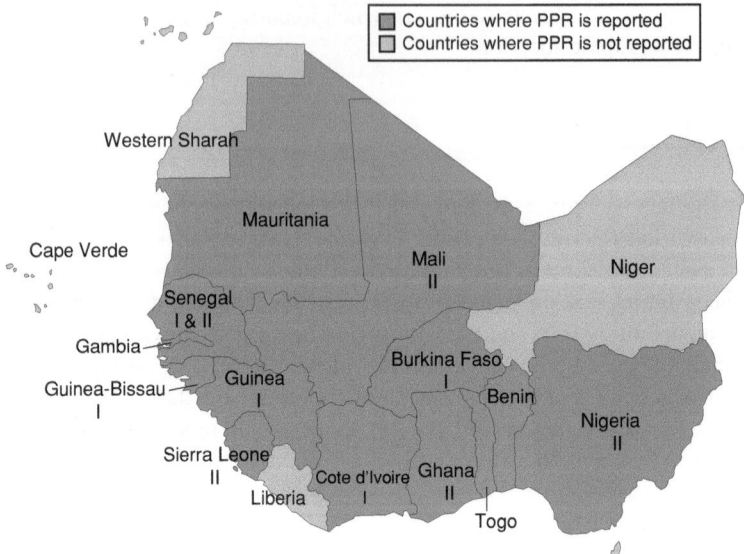

**Fig. 5.5**  Epidemiology and distribution of PPRV in West Africa

region of Turkey (Albayrak and Alkan 2009). Collectively, outbreaks have occurred in both the Anatolia (Asian Turkey) and Thrace (European Turkey) regions. Molecular typing indicates that the Turkish isolates of PPRV belong to lineage IV.

### 5.3.3  Distribution of PPRV in West Africa

West Africa consists of 16 countries in the westernmost region of the African continent, covering an area of approximately 5 million square kilometres. The countries include Benin, Burkina Faso, Cape Verde, Cote d'Ivoire, Gambia, Ghana, Guinea, Guinea-Bissau, Liberia, Mali, Mauritania, Niger, Nigeria, Senegal, Sierra Leone and Togo, besides British overseas territories and South Atlantic Ocean islands. PPRV is reported from all of the countries of West Africa except Niger, West Sahara and Liberia, where PPRV has probably never been investigated so far (Fig. 5.5).

West Africa is the place of origin for PPRV. Initially, a severe disease was observed in sheep and goats in the early 1940s in Cote d'Ivoire, which was not transmissible to large ruminants since in-contact cattle did not show clinical disease (Gargadennec and Lalanne 1942). They first called this condition bluetongue (1940), and then ulcerative stomatitis (1941). In 1942, they finally named it the "peste des petits ruminants" because of clinical similarities with RP. At that time it was also suspected that it is a strain of rinderpest that is adapted to small ruminants.

RP was also a most prevalent disease in the region, and small ruminants were susceptible to RP and therefore were serologically positive. It was impossible to discriminate clinical disease in sheep and goats from RP at that time. It was not until 1979 that Gibbs and others defined PPRV as a distinct entity (Gibbs et al. 1979). After its first report, the disease was reported from other countries in Western African. In early 1955, sheep from Casamance and goats in the Kaolack regions of Senegal showed clinical signs of PPRV, and gave the investigators an opportunity to have a full clinical picture of the condition in its various forms (Mornet et al. 1956).

The first report of PPRV in Nigeria, recognized as stomatitis pneumonitis complex or stomatitis and enteritis of goats, date back to 1967 (Hamdy and Dardiri 1976; Whitney et al. 1967; Mann et al. 1974). Most of the disease description was based on clinical observation and diagnosis by AGID, which were proposed to be able to discriminate PPRV from RPV. Later in 1976, Hamdey et al. (1976) confirmed that PPRV is the cause of the stomatitis pneumonitis complex (Hamdy et al. 1976). The early isolation of PPRV from Nigeria became the prototype, and it was extensively used for experimental studies and as a candidate for vaccine production. This isolate is the most widely used vaccine virus for PPRV around the globe.

Currently, the disease has been reported in all most all the countries of West African (Table 5.1). However, all countries have not reported the existence of the clinical disease: for some reports there is only serologic evidence of infection. Based on the current reports of serologic or nucleic acid detection of PPRV, it is clear that the disease has remained prevalent in the West African countries. The disease has been reported from Burkina Faso in 2008, Ghana in 2010, Nigeria in 2007 and Senegal in 2010. PPRV strains from both lineages I and II are currently circulating across West Africa, although undoubtedly many outbreaks are not characterized at the molecular level. Other cases of PPRV in sheep, goat and camel populations have also recently been described in Nigeria (El-Yuguda et al. 2010); and a further Nigerian study used hemagglutinin tests with faecal material to detect PPRV excretion, and suggested that healthy animals may serve as carriers for PPRV (Obidike et al. 2006). In Burkina Faso, an antibody prevalence to PPRV of 28.5 % has been reported in the north (Sow et al. 2008).

Currently, during a training mission organized at Teko Central Veterinary Laboratory, Makeni, Sierra Leone, we collected samples from goats ($n = 9$) and sheep ($n = 1$) from two smallholders with suspected outbreaks of PPR. After serology tested positive with cELISA, RT-PCR specific for the N gene detected the PPR viruses. The molecular characterisation indicated that the isolates clustered in lineage II with viruses from Mali, Nigeria, and Ghana, and could further be distinguished into two clusters. One virus from Kabala, Sierra Leone, clustered closely with viruses from Mali (Mali/99/1), whereas all others showed 100 % identity with a virus from Nigeria (Nig/75/1), which is the one used as the vaccine virus strain (Munir et al. 2012c). Since this study, an official vaccination program based on Nigeria/75/1 has been launched.

**Table 5.1** Information on the first report of PPR from different countries

| Sr. no | Country | Year of first report | Host | Mode of diagnosis: clinical (C), serology (S) genome (G) | Reference |
|---|---|---|---|---|---|
| 1 | Ivory coast | 1942 | Goats | C | Gargadennec and Lalanne (1942) |
| 2 | Senegal | 1955 | Sheep and goats | C | Mornet et al. (1956) |
| 3 | Nigeria | 1967 | Sheep and goats | C and S | Hamdy et al. (1976), Whitney et al. (1967) |
| 4 | Chad | 1971 | Goats | C | Provost et al. (1972) |
| 5 | Sudan | 1971 | Sheep and goats | C and S | Ali and Taylor (1984) |
| 6 | Togo | 1972 | Sheep and goat | C | Benazet et al. (1973) |
| 7 | Benin | 1972 | Sheep and goats | C | Bourdin (1973) |
| 8 | Oman | 1978 | Sheep and goats | C and S | Hedger et al. (1980) |
| 9 | Saudi Arabia | 1980 | Sheep, goats, deer, gazelles | C and S | Hafez et al. (1987); Asmar et al. (1980) |
| 10 | UAE | 1983 | Gazelles, Ibex, sheep, gemsbok and Nilgai | C and S | Furley et al. (1987) |
| 11 | India | 1987 | Sheep | C and S | Shaila et al. (1989) |
| 12 | Egypt | 1987 | Goats | C and S | Ismail and House (1990) |
| 13 | Pakistan | 1991 | Goats | S and G | Amjad et al. (1996) |
| 14 | Israel | 1993 | NA | NA | Perl et al. (1994) |
| 15 | Bangladesh | 1993 | Goats | S and G | Islam et al. (2001) |
| 16 | Ethiopia | 1994 | Sheep and goats | C and S | Roeder et al. (1994) |
| 17 | Eritrea | 1994 | Sheep and goats | C and S | Anonymous (1994) |
| 18 | Iran | 1995 | Sheep and goats | S | Described in Bazarghani et al. (2006) |
| 19 | Afghanistan | 1995[a] | Sheep and goats | S | Officially described in FAO by Martin and Larfaoui (2003) |
| 20 | Nepal | 1995 | Sheep and goats | S and G | Described in Dhar et al. (2002) |
| 21 | Uganda | 1995 | Goats | C and S | Wamwayi et al. (1995) |
|  | Kenya | 1995 | Goats | C and S | Wamwayi et al. (1995) |
| 22 | Kazakhstan | 1997 | Cattle, sheep and goats | S | Lundervold et al. (2004) |
| 23 | Iraq | 1998 | Sheep | S | FAO (2003) |
| 24 | Vietnam | 2007–2008 | Goats, cattle, buffalo | S | Maillard et al. (2008) |
| 25 | Tajikistan | 2004 | Sheep and goats | S and G | Kwiatek et al. (2007) |
| 26 | Kenya | 2006 | Sheep and goats | C and S | Anonymous (2008) |

(continued)

**Table 5.1** (continued)

| Sr. no | Country | Year of first report | Host | Mode of diagnosis: clinical (C), serology (S) genome (G) | Reference |
|---|---|---|---|---|---|
| 27 | Somalia | 2006 | Sheep and goats | C and S | Anonymous (2008) Nyamweya et al. (2008) |
| 28 | Uganda | 2007 | Sheep and goats | S and C | RO-CEA (2008) |
| 29 | China | 2007 | Sheep and goats | S and G | Wang et al. (2009) |
| 30 | Morocco | 2008 | Sheep and goats | C and S | Sanz-Alvarez et al. (2009) |
| 31 | Tanzania | 2008 | Sheep and goats | C and S | Swai et al. (2009) FEWSNET (2008) |
| 32 | Serra Leone | 2009 | Sheep and goats | S and G | Munir et al. (2012a, b, c) |
| 33 | Algeria | 2011 | Sheep and goats | C and S | OIE (2011) |
| 34 | Tunisia | 2011 | Sheep and goats | C and S | OIE (2011) |

[a] Disease is prevalent since 1995, however there is no official record of any outbreak until 2003. *NA* not available

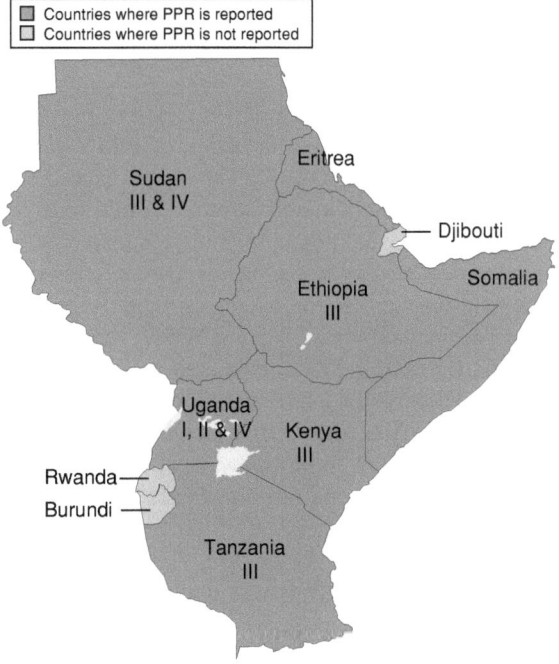

**Fig. 5.6** Epidemiology and distribution of PPRV in East Africa

Countries where PPR is reported
Countries where PPR is not reported

Sudan
III & IV

Eritrea

Djibouti

Ethiopia
III

Somalia

Uganda
I, II & IV

Kenya
III

Rwanda

Burundi

Tanzania
III

## 5.3.4  Distribution of PPRV in East Africa

East or Eastern Africa consists of ten countries, including Tanzania, Burundi, Rwanda, Uganda, Sudan, Ethiopia, Eritrea, Djibouti, Somalia and Kenya (EFC EoFaC 2003). However, Burundi and Rwanda are sometimes considered part of Central Africa. PPRV has been reported in all of the countries of East Africa except Djibouti, Brundi and Rwanda (Fig. 5.6), where it has probably never been investigated. There are informal reports of PPRV in camels in Djibouti [described in (Roger et al. 2000)]; however, no further information has appeared.

### 5.3.4.1  Eritrea

PPRV has been reported in Eritrea since 1994; however, only a clinical description of the disease was available and diagnosis was based on the serologic tests, which unfortunately cannot differentiate PPRV from RPV (Anonymous 1994). Latter in 1998, a series of outbreaks were recorded in the east of the Asmara region in Eritrea, and the viral antigen was confirmed in the conjunctival epithelial cells using immunofluorescent antibody test (IFA), in which monoclonal antibodies specifically raised against PPRV were used. Additionally, the presence of syncytia, a marker for morbilliviruses infection, was detected in conjunctival smears using Giemsa-staining (Sumption et al. 1998).

### 5.3.4.2  Uganda

In Uganda, the serologic detection of PPRV was reported in 1995. However, in March 2007 the Uganda Ministry of Agriculture Animal Industry and Fisheries (MAAIF) recorded the first outbreak of PPRV. It was estimated that 17 % of goats and sheep in the Karamoja region are infected with PPR, posing a severe threat to small ruminants due to the heavy economic losses to farmers as well as the small ruminant industry. In the following year (August 2008), 700,000 goats and sheep were vaccinated by FAO against PPRV, to avoid food and nutritional shortages in Uganda (RO-CEA 2008). Two studies conducted by the same group, with over-lapping target regions, have reported the seroprevalence of PPRV in several of the Ugandan districts (Moroto, Nakapiripirit, Abim and Kotido) with an overall prevalence of 57.6 % (Mulindwa et al. 2011; Luka et al. 2011). In another report it was concluded that the seroprevalence of PPRV among vaccinated, unvaccinated or with unknown vaccination status small ruminants was found to be 55.3 % (84/152), 11.7 % (2/17) or 53.3 % (80/150), respectively in selected districts. Recently, genetic analysis was performed on oculo-nasal and blood samples collected from sheep and goats in the Karamoja region from suspected outbreaks during 2007–2008. Interestingly, the phylogenetic analysis of the F gene of PPRV showed that some isolates (Ugn/14/09; Ugn/16/09; Ugn/18/09) clustered with the Nigerian sequences in lineage I, whereas two of the isolates (Ugn/LF1/07;

Ugn/LF3/09) clustered with Cote d'Ivoire (ICV/86) in lineage II, and one isolate (Ugn/FF/09) clustered with Asia isolates in lineage IV. These findings suggest that heterogeneous strains of West African and Asian lineages are in circulation in the Karamoja region of Uganda. It was speculated that unrestricted and uncontrolled movements of small ruminants within neighbouring countries might cause introduction of multiple African lineages in Uganda (Luka et al. 2012).

### 5.3.4.3  Sudan

Initially, a rinderpest-like disease was observed in three areas in south Gedarif (Eastern Sudan) during 1971 and 1972, and due to cross reactivity and the ability of the RP serum to neutralize the suspected virus, it was concluded to be RP (Hag and Ali 1973). However, later virus isolation from the same samples and its ability to cause disease in sheep and goats confirmed that the causative agent was PPRV (Hag et al. 1984). The disease was then reported from the Sinnar area in central Sudan during 1971–1972 (Rasheed 1992) and in Mieliq areas in western Sudan in 1972, from goats and sheep, respectively (Hassan et al. 1994). After these reports, the disease remained endemic in Sudan, as can be realized from the disease description in sheep and goats in Khartoum State (Zeidan 1994; El Amin and Hassan 1998), PPRV outbreaks in different regions of Sudan [Gezira State, White Nile State (Central), Khartoum state, North Kordofan State (Western) and River Nile State (Northern)] during 2000–2002 (Intisar 2002). Later, a serologic survey conducted during 2002–2005 indicated that the disease is prevalent in Kordofan state and Darfur State with a rate of 70 % and 52.5 %, respectively (Intisar et al. 2007). Besides other reports of PPRV in Sudan, a significant contribution was made by Khalafalla et al. (2010) when they isolated the virus from a camel population. The animals were suffering from severe disease mainly characterized by colic, breathing problems, bloody diarrhoea and abortion (Khalafalla et al. 2010), resembling a previous case reported in Ethiopia during 1995–1996 (Roger et al. 2001a).

In order to understand the current status of small ruminants and camel for PPRV, Saeed et al. (2010) analysed 1198 serum samples collected from sheep ($n = 500$), camels ($n = 392$) and goats ($n = 306$) from different areas in Sudan (Khartoum, Gezira, Tambool, River Nile, Kordofan, White Nile, Blue Nile, Gedarif, Kassala, Halfa ElGadida, Port Sudan). The results of cELISA, which detects the PPRV specific antibodies, demonstrated that the disease in sheep, goats and camel is prevalent with a rate of 67.2 %, 55.6 % and 0.3 %, respectively (Saeed et al. 2010). It has previously been identified that lineage III is the most prevalent type of PPRV (El Hag Ali and Taylor 1984). Recently, in contrast to the expected outcome from Sudan, analysis of the PPRV isolates from 2000–2009 indicated that most of the PPRV strains belong to lineage IV while very few remained in lineage III. Genetic characterisation of both the N and F genes has shown that an Asian lineage IV is being introduced and spread to Africa, and in parallel African lineage III in Sudan is decreasing dramatically. Collectively, the

Sudanese isolates can be subdivided into two sublineages, which belong to either PPRV strains from Saudi Arabia or central Africa (Kwiatek et al. 2011).

### 5.3.4.4  Tanzania

Although the disease has been characterized, both genetically and serologically, in many neighbouring countries, a serologic survey in 1998 in Tanzania did not detect any antibodies to PPRV, which suggested at the time of investigation that infection remained restricted (Wambura 2000). However, given the highly infectious nature of PPRV, it has later been serologically reported in Tanzania (Swai et al. 2009). In this report, serologic analysis was conducted in sheep and goat flocks from seven different geographic administration authorities (Ngorongoro, Monduli, Longido, Karatu, Mbulu, Siha and Simanjiro) located in Northern Tanzania, because of their close proximity to Southern Kenya, where PPR has been reported to have decimated small ruminants in the recent past (FEWSNET 2008). The results of cELISA performed on sera collected from 657 sheep and 892 goats indicated a high sero-prevalence (45.8 %) (Swai et al. 2009). This confirmed outbreak threatens over 13.5 million goats and 3.5 million sheep in the country. Later, a study was conducted by Kivaria et al. (2009) to determine the seroprevalence, distribution, isolation and characterisation of an emerging PPRV infection in sheep and goats in Tanzania. A total of 1546 serum samples from small ruminants reared in 48 villages from the 7 districts were investigated. It was observed that the prevalence of PPRV infection varied (range 0.0–14.00 %) and was higher in goats (50 %) than in sheep (40 %). The overall antibody response to PPRV was 45.0 % (Kivaria et al. 2009). Preliminary diagnosis of PPR from cases in the Tandahimba district of Mtwara region has indicated the presence of PPR in Mtwara (Girald Misinzo, Sokoine University of Agriculture, Tanzania, personal communication). Based on these reports, FAO-EMPRESS gave an alert on the suspected outbreak of PPR in Southern Tanzania.

Although comprehensive knowledge of the genetic makeup of PPRV in Tanzania is currently lacking, characterisation of PPRV from sheep blood and tissues samples from Tanzania showed that the isolated strains belong to lineage III and are closely related to the PPRV isolated from East Africa and the Middle East. Thus, one of the neighbouring countries in the Eastern Africa region is most likely the source of infection (Kivaria et al. 2009). The presence of PPR in the South Zone poses a risk of the disease spreading further South to whole of 15-nation Southern African Development Community (SADC). Therefore, an emergency vaccination programme has been launched and implemented in the northern half of Tanzania. Additionally, it was recommended by the FAO that vaccination should be considered in the area bordering Malawi, Mozambique and Zambia. It was also advised that these countries step up their vigilance for the disease and conduct proactive surveillance (FAO 2010).

### 5.3.4.5 Kenya

Kenya is among those countries where PPRV has recently been confirmed. A PPR-like disease appeared in 2006 in the Oropoi and Lokichoggio divisions of the Turkana District in Kenya. This report was followed by disease appearance in 16 other districts in the North Rift region of Kenya, in which Samburu West, Samburu East, Pokot, Marakwet, Baringo and Keiyo districts remained the most affected. The disease has since affected livestock in 46 districts in Kenya in the North Eastern, Eastern and Coast Provinces (Nyamweya et al. 2008). These reports of PPRV have left strong socioeconomic consequences for food security and have impacted negatively on the livelihoods of the local population. In 2008, IRIN reported that in a severe PPRV outbreak 300 goats from a herd of 800 were succumbed in just three months in the Turkana region of North-western Kenya (IRIN 2008). At this time Morris Lichokwe, a community leader at Kenya, said:

> *Lomoo (PPRV) has really brought us down*

During 2006 to 2008, more than 5 million animals were affected in as many as 16 districts in Kenya, and 2.5 million died of PPRV (Anonymous 2008). Having these devastating outcomes of PPRV, in already drought and clan clashing areas, vaccination and quarantine have been used to stop the continued spread of PPRV in Kenya. However, inadequate funding, limited stocks of available vaccine, shortage of trained staff to coordinate vaccination programmes, tribal clashes, drought and the mobility of the pastoral communities involved have made the task more problematic (Anonymous 2008). The genetic nature of PPRV circulating in Kenya is currently not known; however, it is likely that Kenyan PPRV strains may belong to lineage III, due to the history of such lineage in other neighbouring countries.

### 5.3.4.6 Somalia

The disease was initially reported in the central regions of Hiran, Middle Shabelle and Galgadud in Somalia, at the same time as in Kenya (during 2006) (Nyamweya et al. 2008; Anonymous 2008). Because of this initial report, the collaborative efforts made by Somalia Animal Health Services Project (SAHSP) and Ministry of Livestock, Forestry and Range in the Hiran region provided substantial contribution in the confirming of the disease. Based on follow up studies, they concluded that, because of the geographic position of Somalia, the disease outbreaks remained restricted, and other parts of Somalia remained free from PPRV. Additionally, with the collaborative efforts of SAHSP, Cooperazione Internationale (COOPI) and Vétérinaires Sans Frontières Suisse (VSF-S), ring vaccination was recommended to contain the spread of disease to the surrounding region.

### 5.3.4.7 Ethiopia

In Ethiopia, PPR was clinically suspected for the first time in 1977 in a goat herd near the Afar region, in the East of Ethiopia (Pegram and Tereke 1981). However, it was not until 1984 that Taylor (1984) observed the disease clinically and provided evidence serologically. This was later, for the first time, confirmed by Roeder et al. (1994) when 60 % of a goat herd ($n = 1432$ animals) succumbed in the south of the Addis Ababa region. Primary diagnosis was made based on AGID, followed by analysis made by virus-specific cDNA probes against the N protein of PPRV, and confirmation was made by cELISA. Based on the results, predicted by all of the assays, the investigators were able to conclude that this high mortality was due to PPRV and not to RPV, which was the most prevalent disease in cattle at that time (Roeder et al. 1994). To further assess the status of PPRV in selected urban areas, Roger and Bereket (CIRAD-EMVT report n°96006, Montpellier, 1996) reported a seroprevalence of 33 % and 67 % in sheep and goats, respectively. Serologic reports afterward gave somewhat biased frequency and distribution of PPRV in Ethiopia (Gelagay 1996; Abraham et al. 2005). However, it become clear in the late 1990 s that the virus has been circulating extensively among the small ruminant population of Ethiopia, and it is likely that PPR will become, if not already, one of the most economically important livestock diseases in the country (Gopilo 2005). Currently, results of Waret-Szkuta et al. (2008), based on 13,651 serum samples collected from small ruminants, indicated that PPRV circulation is very heterogeneous in Ethiopia, and there is large variation between regions and *weredas* (composition of *kebelles* or *Peasant Associations* that are an aggregation of *got*, a *got* being a group of three to five villages) (Waret-Szkuta et al. 2008).

Until 1992, PPRV was considered a disease only of small ruminants. Since then, several reports demonstrated the PPRV antibodies in camels, providing evidence that camels are also victims of PPRV (Ismail et al. 1992; Haroun et al. 2002; Abraham et al. 2005; Albayrak and Gur 2010). It was in Ethiopia that PPRV was first reported and confirmed in camels, where it caused highly contagious respiratory syndromes with high illness rates but low death rates (Roger et al. 2001b). Genetic typing of the Ethiopian PPRV isolates indicated that these belong to lineage III. Interestingly, surveillance of camels revealed consecutive outbreaks in: Kassala, eastern Sudan (2004); Atbara, northern Sudan (2005); and Tambool, Blue Nile region, Sudan (2007) (Khalafalla et al. 2010). Genetic analysis of these isolates showed a close relationship to the viruses isolated from sheep and goats in the same regions, providing clues that camels may have served as a bridge with areas of northern Africa and contributed to the spread of a camel-derived strain of lineage IV, as seen in Morocco. However, such an interpretation requires rigorous research to be conclusive. Moreover, the role of camels in the epizootiology of the disease still remains to be determined.

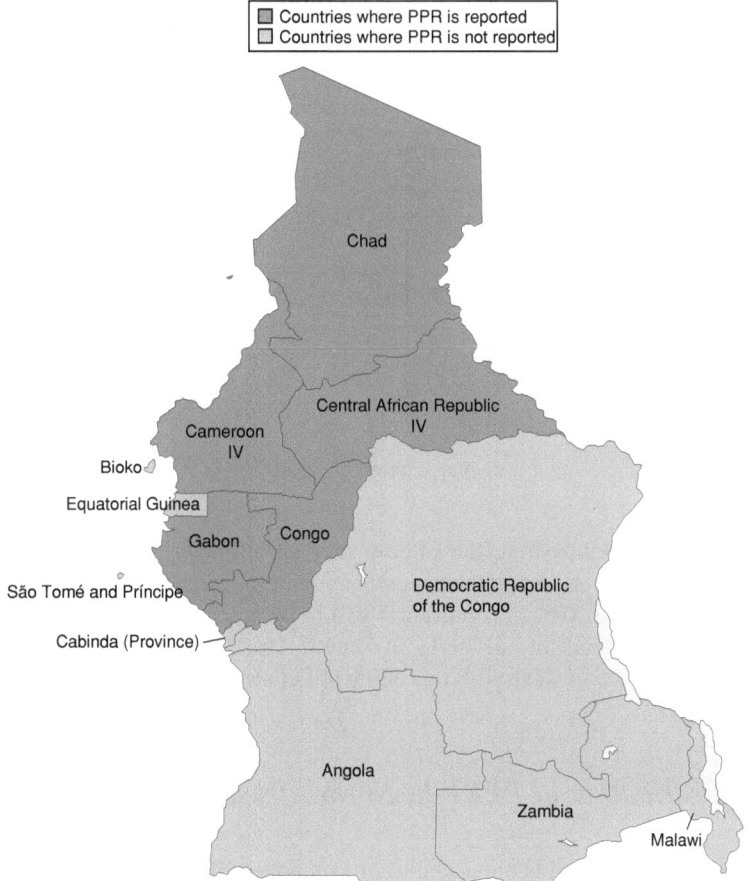

**Fig. 5.7** Epidemiology and distribution of PPRV in Central Africa

## 5.3.5 *Distribution of PPRV in Central Africa*

Central African countries include Angola, Cameroon, the Central African Republic, Chad, the Republic of the Congo, the Democratic Republic of the Congo, Equatorial Guinea, Gabon, and São Tomé and Príncipe. Most of the reports from Central African countries are based on either clinical assessment or serologic demonstration (Fig. 5.7). These reports have demonstrated the disease in the Central African Republic during 1999, 2005 and 2006, the Democratic Republic of the Congo in 2006, Chad in 1999 and 2006, and recently in Cameroon (Awa et al. 2002) and Gabon in 2007 (Banyard et al. 2010). The first report of PPRV in Chad dates back to the 1980s (Provost et al. 1972). In early 1993, a serologic prevalence of 34 % was observed using an ELISA test in Sahelian goats, and the isolated viruses were used for experimental inoculation in goats. Serologic tests such as AGID and ELISAs

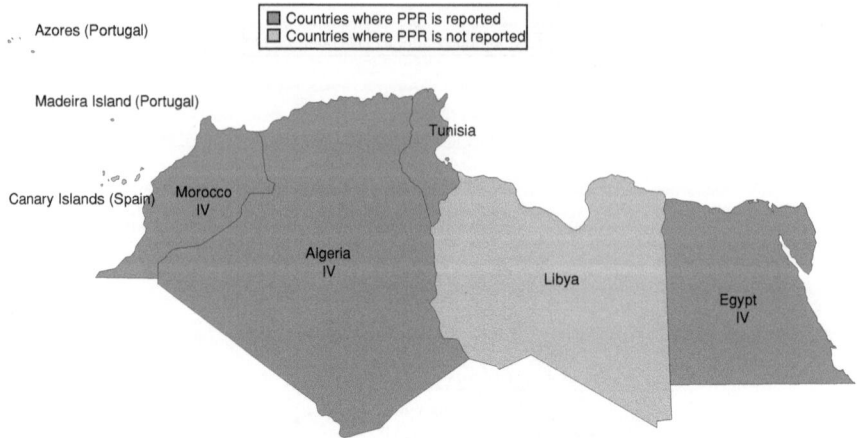

**Fig. 5.8**  Epidemiology and distribution of PPRV in North Africa

were applied to demonstrate the disease predisposition in the inoculated flocks (Bidjeh et al. 1995). Although the genetic nature of the circulating PPRV viruses in some of the Central African countries is still to be evaluated, but it is plausible to consider the prevalence of lineage IV in Central Africa, due to confirmation of this lineage in countries of Central Africa (Banyard et al. 2010).

## 5.3.6  Distribution of PPRV in North Africa

According to the United Nations definition, geopolitically North Africa consists of Algeria, Egypt, Libya, Morocco, Tunisia, Sudan and Western Sahara. The last two countries are also considered to be part of West Africa, and Algeria, Morocco, Tunisia, Mauritania, and Libya together are referred to as the Maghreb or Maghrib. North Africa also includes a number of Spanish and Portuguese islands. The location of North Africa is peculiar due to Egypt, which is a transcontinental country by virtue of the Sinai Peninsula, which is in Asia. Additionally, North Africa is historically and ecologically distinct because of the effective barrier created by the Sahara, and this geographic division is considered crucial in disease ecology. Among the countries in North Africa, PPRV has been reported from Egypt, Morocco, Algeria, and Tunisia, whereas it is only Libya where disease has not been reported to date (Fig. 5.8).

### 5.3.6.1  Egypt

In Egypt, PPRV was first described in 1987, when a rinderpest like disease appeared in goats causing high mortality (30 %), and morbidity (90 %) (Ismail and House 1990). After this initial and preliminary report of PPRV, the same group isolated and infected Egyptian goats and Boscat rabbits with the Egyptian strain of PPRV (Egypt 87). Although the disease didn't spread by contact, the virus caused disease with minor clinical manifestations accompanied by a detectable level of neutralizing antibodies (Ismail et al. 1990). Recently, a study conducted by El-Hakim, (2006) demonstrated that the disease remained in the Aswan province where some goats showed severe clinical disease while others remained asymptomatic (El-Hakim 2006). Genetic characterisation of the F gene of PPRV revealed that the Egyptian strain belongs to lineage IV and is closely related to recently isolated Turkish strains of PPRV.

### 5.3.6.2 Morocco

Outbreaks of PPRV in Morocco raised immense concerns at the severity of the disease, because it had been suggested that PPRV in Egypt is due to its trans-continental status and that the remainder of North Africa is totally free from PPRV. The first outbreak of PPRV in Morocco occurred on 12 June 2008 in the rural village of Ain Chkef, Moulay Yacoub, which is close to Fés (Sanz-Alvarez et al. 2009). OIE confirmed on 23 July 2008 that these outbreaks were caused by PPRV. Until August, there have been several outbreaks of PPRV throughout the country including the reports from the border of Morocco and Algeria. Collectively, 257 outbreaks were recorded in 36 out of 61 total provinces of Morocco. Due to this devastating wave of PPRV, a mass vaccination programme was implemented, in which approximately 20.6 million of Morocco's sheep and goat population were vaccinated. Although, the origin of PPRV outbreaks in Morocco was not ruled out, it was suspected that the introduction occur due to movement of the live but infected animals, since there are intense migratory movements of Saharan nomad populations at the border of North African countries. Later, genetic characterisation of the Moroccan virus classified it in lineage IV (Khalafalla et al. 2010). Recently, analysis of the sheep sample collected during the 2008 outbreaks indicated that lineage IV is present in Morocco, and that these PPRV strains are closely related to Saudi Arabian strains and show a difference of only four nucleotides (Kwiatek et al. 2011). Recently, Mikael Baron's group at IAH, UK, have sequenced the complete genome of the Moroccan PPRV isolate (S. Parida personal communication).

### 5.3.6.3 Algeria

Although PPRV positive small ruminants were identified in 2008 at the border of Morocco adjacent to Algeria, it was not until recently when serologic evidence was officially reported from Algeria for the first time. Before the first official report of PPRV to OIE in 2011, the seropositive small ruminants have been observed in 2005 and 2008 in western Algeria [Broglia et al. unpublished data, described in (De Nardi et al. 2012)]. In February 2011, seven outbreaks of a subclinical disease were reported in five provinces (Naama, Bechar, Adrar, Tamanrasset, and

Tindouf) of the southwestern part of Algeria. The animals appeared to be seropositive as measured by cELISA; however, they tested negative with RT-PCR (OIE 2011a) which made the prediction of the source of virus origin difficult. Recently, De Nardi et al. (2011) implemented a survey in the Sahrawi refugee camps (western Algeria) in May 2010, which resulted in the detection of PPRV genetic material in three of nine sampled animals (De Nardi et al. 2012). Molecular typing and phylogenetic analysis characterized the strain as belonging to lineage IV. The phylogenetic analysis indicated a close relationship with the PPRV isolated during the Moroccan PPR outbreak in 2008. Although the origin of the outbreak remains unknown, there was a history of importation of small ruminants from Mauritania and the liberated territories of Western Sahara for the celebration of the "Eid-ul-Adha" festival during 2009. It was also speculated that animals could have been illegally moved directly from Morocco or from the Moroccan Southern Province into Algeria [Lamin Saleh, personal communication described in (De Nardi et al. 2012)].

#### 5.3.6.4 Tunisia

OIE has recently presented serologic evidence for PPRV infection in Tunisia (OIE 2011b). Later, samples collected from 263 sheep and 119 goats from six regions (Bizerte, Kairouan, Kébili, Médenine, Nabeul and Sousse) of Tunisia were found to be serologically positive by cELISA. However, 28 lung samples harvested from slaughtered animals ($n = 25$ sheep and $n = 3$ goats) from different regions were tested negative using a PPR virus-specific RT-PCR (Ayari-Fakhfakh et al. 2011). There is great ongoing movement of livestock between Algeria and Libya, which poses a risk of PPRV spread to Libya, which is the only country in North Africa that is PPRV free. However, the virus may well be present across other, as yet unknown, regions of Libya.

### 5.4 PPRV and Europe

Although the disease has been circulating in sub-Saharan Africa for several decades and in the Middle East and Southern Asia since 1993, the occurrence of new outbreaks in previously PPR free countries such as Morocco is an alarming situation for the neighbouring countries. Algeria, having 19 million sheep and 3 millions goats, is highly vulnerable to be affected being close to Morocco; and serologic evidence has indicated the presence of the disease in Algeria.

This also poses a serious threat for introduction of PPR in European Union countries, notably Spain, given its geographic location and a susceptible population of 23 million sheep and 3 million goats (FAO 2008, September 9). Historical exchanges exist between Morocco and Spain, where both ovine and caprine populations are important. Moreover, increase in human population and in turn the small ruminant population across such areas poses the risk of further emergence of

PPRV across North Africa. Additionally, outbreaks have occurred in both Anatolia (Asian Turkey) and Thrace (European Turkey), which has raised the concern of PPRV in Eastern Europe, given the immense trade between Turkey and some of the European Union countries.

Since the first recent report of outbreaks in Morocco and Turkey, the situation regarding PPRV in and around Europe has changed dramatically, with incursions of the virus in previously disease-free area, affecting immunological naïve herds and generating important economical losses. It is therefore essential that Europe maintain surveillance of the disease in order to successfully contain the disease.

## 5.5   Conclusions

There have been immense efforts to determine and characterize the best classification systems for PPRV. Lineage IV was originally considered an Asian group of PPRV. However, recently it is considered that lineage IV is overwhelming the other lineages in African countries while still being predominant in Asia. In general, most of the recent reports of PPRV in previously PPRV free countries belong to lineage IV, which suggests that lineage IV is a novel group of PPRV and may replace the other lineages in the near future.

Another aspect of PPRV origin is that, despite being first identified in West Africa, the disease appeared in India in the early 1940s, from where it spread to West Africa (Taylor and Barrett 2010). Parallel comparisons of the disease reports show that the disease appeared in Pakistan, Iran, Iraq, and Bangladesh at the same time (1993–1998), further supporting the disease origin in the Indian Subcontinent.

At one time, PPR was thought to be restricted to Western Africa, but it has since been recognized from the equator line up to the Sahara desert, as well as in Asia and the Middle East. Other nearby areas, such as Southern Africa and Central Asia, are under increasing threat of disease dissemination. Epidemic outbreaks in previously non-infected countries have been associated with severe consequences on livelihoods, and are of great concern to countries neighbouring the newly infected countries. Burundi, Rwanda and other southern African countries are at risk. It is still unclear whether differences between lineages merely reflect geographic speciation or if they are also correlated with variability in pathogenicity between isolates. Collectively, recognition of PPRV in larger economies such as China, Turkey, India and Pakistan will likely make substantial contribution toward understanding of the disease distribution pattern and ecology of PPRV.

# References

Abraham G, Sintayehu A, Libeau G, Albina E, Roger F, Laekemariam Y, Abayneh D, Awoke KM (2005) Antibody seroprevalences against peste des petits ruminants (PPR) virus in camels, cattle, goats and sheep in Ethiopia. Prev Vet Med 70(1–2):51–57

Abu Elzein EM, Hassanien MM, Al-Afaleq AI, Abd Elhadi MA, Housawi FM (1990) Isolation of peste des petits ruminants from goats in Saudi Arabia. Vet Rec 127(12):309–310

Abubakar M, Jamal SM, Hussain M, Ali Q (2008) Incidence of peste des petits ruminants (PPR) virus in sheep and goat as detected by immuno-capture ELISA (Ic ELISA). Small Rumin Res 75(2):256–259

Al-Dubaib MA (2008) Prevalence of peste des petits ruminants virus infection in sheep and goat farms at the central region of Saudi Arabia. Res J Vet Sci 1(1):67–70

Albayrak H, Alkan F (2009) PPR virus infection of sheep in black sea region of Turkey: epidemiology and diagnosis by RT-PCR and virus isolation. Vet Res Commun 33:241–249

Albayrak H, Gur S (2010) A serologic investigation for peste des petits ruminants infection in sheep, cattle and camels (*Camelus dromedarius*) in Aydin province, West Anatolia. Trop Anim Health Prod 42(2):151–153

Alcigir G, Vural SA, Toplu N, kuzularda Turkiye'de (1996) Peste des petits ruminants virus enfeksiyonunun patomorfolojik ve immunohistolojik ilk tanimi. Ankara Universitesi Veteriner Fakültesi Dergisi 43:181–189

Ali EHB, Taylor WP (1984) Isolation of peste des petits ruminants virus from the Sudan. Res Vet Sci 36(1):1–4

Amjad H, Qamar-ul I, Forsyth M, Barrett T, Rossiter PB (1996) Peste des petits ruminants in goats in Pakistan. Vet Rec 139(5):118–119

Anonymous (1994) Animal health yearbook 1993, Food and Agriculture Organization of the United Nations (FAO). http://www.amazon.co.uk/Animal-Health-Yearbook-1993-Production/dp/9250035276. Accessed 22 May 2010

Anonymous (2008) Kenya: conflict and drought hindering livestock disease control

Arzt J, White WR, Thomsen BV, Brown CC (2010) Agricultural diseases on the move early in the third millennium. Vet Pathol 47(1):15–27

Asmar JA, Radwan AI, Abi Assi N, Al-Rashied A (1980) PPR-like disease in sheep of central Saudi Arabia: evidence of its immunological relation to rinderpest; prospects for a control method. Paper presented at the 4th annual meeting of the Saudi biological society, University of Riyadh Press, Riyadh, 10–13 March

Attieh E (2007) Enquete sero-epidemiologique sur les principales maladies caprines au Liban. Ecole Nationale Veterinaire de Toulouse–ENVT. http://oataouniv-toulousefr/1812/

Awa DN, Ngagnou A, Tefiang E, Yaya D, Njoya A (2002) Post vaccination and colostral peste des petits ruminants antibody dynamics in research flocks of Kirdi goats and Foulbe sheep of North Cameroon. Prev Vet Med 55(4):265–271

Ayari-Fakhfakh E, Ghram A, Bouattour A, Larbi I, Gribaa-Dridi L, Kwiatek O, Bouloy M, Libeau G, Albina E, Cetre-Sossah C (2011) First serological investigation of peste-des-petits-ruminants and Rift Valley fever in Tunisia. Vet J 187(3):402–404

Balamurugan V, Krishnamoorthy P, Veeregowda BM, Sen A, Rajak KK, Bhanuprakash V, Gajendragad MR, Prabhudas K (2012) Seroprevalence of peste des petits ruminants in cattle and buffaloes from Southern Peninsular India. Trop Anim Health Prod 44(2):301–306

Balamurugan V, Sen A, Venkatesan G, Yadav V, Bhanot V, Riyesh T, Bhanuprakash V, Singh RK (2010) Sequence and phylogenetic analyses of the structural genes of virulent isolates and vaccine strains of peste des petits ruminants virus from India. Transboundary Emerg Dis 57(5):352–364. doi:10.1111/j.1865-1682.2010.01156.x

Banyard AC, Parida S, Batten C, Oura C, Kwiatek O, Libeau G (2010) Global distribution of peste des petits ruminants virus and prospects for improved diagnosis and control. J Gen Virol 91(Pt 12):2885–2897

Bao J, Wang Z, Li L, Wu X, Sang P, Wu G, Ding G, Suo L, Liu C, Wang J, Zhao W, Li J, Qi L (2011) Detection and genetic characterization of peste des petits ruminants virus in free-living bharals (*Pseudois nayaur*) in Tibet, China. Res Vet Sci 90(2):238–240

Barhoom S, Hassan W, Mohammed T (2000) Peste des petits ruminants in sheep in Iraq. Iraqi J Vet Sci 13:381–385

Barrett T (1999) Morbillivirus infections, with special emphasis on morbilliviruses of carnivores. Vet Microbiol 69(1–2):3–13

Bazarghani TT, Charkhkar S, Doroudi J, Bani Hassan E (2006) A review on peste des petits ruminants (PPR) with special reference to PPR in Iran. J Vet Med B Infect Dis Vet Public Health 53(Suppl 1):17–18

Benazet BGH (1973) La peste des petits ruminants: Etude experimentale de la vaccination. C.f. Taylor WP (1984)

Bidjeh K, Bornarel P, Imadine M, Lancelot R (1995) First- time isolation of the peste des petits ruminants (PPR) virus in Chad and experimental induction of the disease. Revue d'elevage et de medecine veterinaire des pays tropicaux 48(4):295–300

Bourdin P (1973) La peste des petits ruminants (PPE) et sa prophylaxie au Senegal et en Afrique de l'ouest. Rev Elev Med Vet Pays Trop 26(4):71a–74a

Chavran V, Digraskar S, Bedarkar S (2009) Seromonitoring of peste des petits ruminants virus (PPR) in goats (*Capra hircus*) of Parbhani region of Maharastra. Vet World 2:299–300

De Nardi M, Lamin Saleh SM, Batten C, Oura C, Di Nardo A, Rossi D (2011) First evidence of peste des petits ruminants (PPR) virus circulation in Algeria (Sahrawi Territories): outbreak investigation and virus lineage identification. Transboundary Emerg Dis 59(3):214–222

Dhar P, Sreenivasa BP, Barrett T, Corteyn M, Singh RP, Bandyopadhyay SK (2002) Recent epidemiology of peste des petits ruminants virus (PPRV). Vet Microbiol 88(2):153–159

Diallo A, Barrett T, Lefevre PC, Taylor WP (1987) Comparison of proteins induced in cells infected with rinderpest and peste des petits ruminants viruses. J Gen Virol 68(Pt 7):2033–2038

EFC EoFaC (2003) East Africa comprises ten countries: Tanzania, Burundi, Rwanda, Uganda, Sudan, Ethiopia, Eritrea, Djibouti, Somalia, and Kenya. Encyclopedia of Food and Culture. The Gage Group Inc

El Amin MAG, Hassan AM (1998) The seromonitoring of rinderpest throughout Africa, phase III results for 1998. IAEA, VINNA, Food and Agriculture Organization/International Atomic Energy Agency, Austria

El Hag Ali B, Taylor WP (1984) Isolation of peste des petits ruminants virus from the Sudan. Res Vet Sci 36(1):1–4

El-Hakim O (2006) An outbreak of peste des petits ruminants virus at Aswan province, Egypt: evaluation of some novel tools for diagnosis of PPR. Assiut Vet Med J 52:146–157

El-Yuguda A, Chabiri L, Adamu F, Baba S (2010) Peste des petits ruminants virus (PPRV) infection among small ruminants slaughtered at the central abattoir, Maiduguri, Nigeria. Sahel J Vet Sci 8:51–62

EMPRES (2000) Emergency prevention system for transboundary plant and animal pests and diseases 2000. EMPRES, 13

Esmaelizad M, Jelokhani-Niaraki S, Kargar-Moakhar R (2011) Phylogenetic analysis of peste des petits ruminants virus (PPRV) isolated in Iran based on partial sequence data from the fusion (F) protein gene. Turkish J Biol 35:45–50

FAO (2003) Pest des petits ruminants in Iraq. FAO corporate document repository. http://www.fao.org/DOCREP/003/X7341E/X7341e01.htm. Accessed 22 May 2010

FAO (2010) Concerns grow about PPR in Tanzania. Vet Rec 167:804

FAO (2008) Outbreak of 'peste des petits ruminants' in Morocco. FAO Newsroom (FAO), Italy

FEWSNET (2008) Livestock disease in Kenya and Uganda worsening food insecurity, threatens to spread

Furley CW, Taylor WP, Obi TU (1987) An outbreak of peste des petits ruminants in a zoological collection. Vet Rec 121(19):443–447

Gargadennec L, Lalanne A (1942) La peste des petits ruminants. Bulletin des Services Zoo Technique et des Epizootie de l'Afrique Occidentale Française 5:16–21

Gelagay A (1996) Epidemiological and serological investigation of multi-factorial ovine respiratory disease and vaccine trial on the high land of North Shewa, Debre Zeit Faculty of Veterinary Medicine, Ethiopia

Gibbs EP, Taylor WP, Lawman MJ, Bryant J (1979) Classification of peste des petits ruminants virus as the fourth member of the genus Morbillivirus. Intervirology 11(5):268–274

Gopilo A (2005) Epidemiology of peste des petits ruminants virus in Ethiopia and molecular studies on virulence. Institut National Polytechnique de Toulouse

Govindarajan R, Koteeswaran A, Venugopalan AT, Shyam G, Shaouna S, Shaila MS, Ramachandran S (1997) Isolation of Pestes des Petits ruminants virus from an outbreak in Indian buffalo (*Bubalus bubalis*). Vet Rec 141(22):573–574

Hafez SM, Sukayran AA, Cruz DD, Bakairi SI, Radwan AI (1987) Serological evidence for the occurrence of PPR among deer and gazelles in Saudi Arabia. Paper presented at the Symposium on the potential of wildlife conservation in Saudi Arabia, National Commission for Wildlife Conservation, Riyadh, 15–18 February

Hag E, Ali B (1973) A natural outbreak of rinderpest involving sheep, goats and cattle in Sudan. Bull Epizoot Dis Afr 12:421–428

Hag E, Ali B, Taylor WP (1984) The isolation of PPR from the Sudan. Res Vet Sci 36:1–4

Hamdy FM, Dardiri AH (1976) Response of white-tailed deer to infection with peste des petits ruminants virus. J Wildl Dis 12(4):516–522

Hamdy FM, Dardiri AH, Nduaka O, Breese SRJ, Ihemelandu EC (1976) Etiology of the stomatitis pneumcenteritis complex in Nigerian dwarf goats. Can J Comp Med 40:276–284

Haroun M, Hajer I, Mukhtar M, Ali BE (2002) Detection of antibodies against peste des petits ruminants virus in sera of cattle, camels, sheep and goats in Sudan. Vet Res Commun 26(7):537–541

Hassan AKM, Ali YO, Hajir BS, Fayza AO, Hadia JA (1994) Observation on epidemiology of peste des petits ruminant in Sudan. Sudan J Vet Res 13:29–34

Hedger RS, Barnett ITR, Gray DF (1980) Some virus diseases of domestic animals in the Sultanate of Oman. Trop Anim Health Prod 12:107–114

Hilan C, Daccache L, Khazaal K, Beaino T, Massoud E, Louis F (2006) Sero-surveillance of "peste des petits Ruminants" PPR in Lebanon. Leban Sci J 7(1):9–24

Hoffmann B, Wiesner H, Maltzan J, Mustefa R, Eschbaumer M, Arif FA, Beer M (2012) Fatalities in wild goats in kurdistan associated with peste des petits ruminants virus. Transboundary Emerg Dis 59(2):173–176

Housawi F, Abu Elzein E, Mohamed G, Gameel A, Al-Afaleq A, Hagazi A, Al-Bishr B (2004) Emergence of peste des petits ruminants virus in sheep and goats in Eastern Saudi Arabia. Revue d'elevage et de medecine veterinaire des pays tropicaux 57:31–34

Intisar KS (2002) Studies on peste des petits ruminants (PPR) disease in Sudan. University of Khartoum, Sudan

Intisar KS, Khalafalla AI, El Hassan SM, El Amin MA (2007) Detection of peste des petits ruminants (PPR) antibodies in goats and sheep in different areas of Sudan using competitive ELISA. In: Proceedings of the 12th international conference of the association of institutions for tropical veterinary medicine, Montpellier, France, p 427

IRIN (2008) KENYA: livestock disease, high prices fuelling food insecurity. IRIN Humanitarian News and Analysis, Lodwar

Islam MR, Shamsuddin M, Rahman MA, Das PM, Dewan ML (2001) An outbreak of peste des petits ruminants in Black Bengal goats in Mymensingh, Bangladesh. Bangladesh Vet 18:14–19

Ismail IM, House J (1990) Evidence of identification of peste des petits ruminants from goats in Egypt. Archiv fur experimentelle Veterinarmedizin 44(3):471–474

Ismail IM, Mohamed F, Aly NM, Allam NM, Hassan HB, Saber MS (1990) Pathogenicity of peste des petits ruminants virus isolated from Egyptian goats in Egypt. Archiv fur experimentelle Veterinarmedizin 44(5):789–792

Ismail TM, Hassas HB, Nawal M, Rakha GM, Abd El-Halim MM, Fatebia MM (1992) Studies on prevalence of rinderpest and peste des petits ruminants antibodies in camel sera in Egypt. Vet Med J Giza 10:49–53

Kataria AK, Kataria N, Gahlot AK (2007) Large scale outbreak of peste des petits ruminants virus in sheep and goats in Thar desert of India. Slovenian Vet Res 44(4):123–132

Kaukarbayevich KZ (2009) Epizootological analysis of PPR spread on African continent and in Asian countries African. J Agric Res 4(9):787–790

Kerur N, Jhala MK, Joshi CG (2008) Genetic characterization of Indian peste des petits ruminants virus (PPRV) by sequencing and phylogenetic analysis of fusion protein and nucleoprotein gene segments. Res Vet Sci 85(1):176–183

Khalafalla AI, Saeed IK, Ali YH, Abdurrahman MB, Kwiatek O, Libeau G, Obeida AA, Abbas Z (2010) An outbreak of peste des petits ruminants (PPR) in camels in the Sudan. Acta Trop 116(2):161–165

Khan HA, Siddique M, Sajjad-ur R, Abubakar M, Ashraf M (2008) The detection of antibody against peste des petits ruminants virus in sheep, goats, cattle and buffaloes. Trop Anim Health Prod 40(7):521–527

Kinne J, Kreutzer R, Kreutzer M, Wernery U, Wohlsein P (2010) Peste des petits ruminants in Arabian wildlife. Epidemiol Infect 138(8):1211–1214. doi:10.1017/S0950268809991592

Kivaria FM, Kwiatek O, Kapaga AM, Geneviève L, Mpelumbe-Ngeleja CAR, Tinuga DK (2009) Serological and virological investigations on an emerging peste des petits Ruminants Virus infection in sheep and goats in Tanzania. Paper presented at the 13th East, Central and Southern African commonwealth veterinary association regional meeting and international scientific conference, Kampala, Uganda, 8–13 November

Kul O, Kabakci N, Atmaca HT, Ozkul A (2007) Natural peste des petits ruminants virus infection: novel pathologic findings resembling other morbillivirus infections. Vet Pathol 44(4):479–486

Kwiatek O, Ali YH, Saeed IK, Khalafalla AI, Mohamed OI, Obeida AA, Abdelrahman MB, Osman HM, Taha KM, Abbas Z, El Harrak M, Lhor Y, Diallo A, Lancelot R, Albina E, Libeau G (2011) Asian lineage of peste des petits ruminants virus, Africa. Emerg Infect Dis 17(7):1223–1231

Kwiatek O, Minet C, Grillet C, Hurard C, Carlsson E, Karimov B, Albina E, Diallo A, Libeau G (2007) Peste des petits ruminants (PPR) outbreak in Tajikistan. J Comp Pathol 136(2–3):111–119

Libeau G, Diallo A, Calvez D, Lefevre PC (1992) A competitive ELISA using anti-N monoclonal antibodies for specific detection of rinderpest antibodies in cattle and small ruminants. Vet Microbiol 31(2–3):147–160

Luka PD, Erume J, Mwiine FN, Ayebazibwe C (2011) Seroprevalence of peste des petits ruminants antibodies in sheep and goats after vaccination in Karamoja, Uganda: implication on control. Int J Anim Vet Adv 3(1):18–22

Luka PD, Erume J, Mwiine FN, Ayebazibwe C (2012) Molecular characterization of peste des petits ruminants virus from the Karamoja region of Uganda (2007–2008). Arch Virol 157(1):29–35

Lundervold M, Milner-Gulland EJ, O'Callaghan CJ, Hamblin C, Corteyn A, Macmillan AP (2004) A serological survey of ruminant livestock in Kazakhstan during post-Soviet transitions in farming and disease control. Acta Vet Scand 45(3–4):211–224

Maillard JC, Van KP, Nguyen T, Van TN, Berthouly C, Libeau G, Kwiatek O (2008) Examples of probable host-pathogen co-adaptation/co-evolution in isolated farmed animal populations in the mountainous regions of North Vietnam. Ann N Y Acad Sci 1149:259–262

Mann E, Isoun TT, Gabiyi A, Odegbo-Olukoya OO (1974) Experimental transmission of the stomatitis pneumonitis complex to sheep and goats. Bull Epizoot Dis Afr 22:99–102

Martin V, Larfaoui F (2003) Suspicion of foot-and-mouth disease (FMD)/peste des petits ruminants (PPR) in Afghanistan (5/05/2003). http://www.fao.org/eims/secretariat/empres/eims_search/1_dett.asp?culling=simple_s_result&publication=&webpage=&photo=&press=&lang=en&pub_id=145377. Accessed 22 May 2010

Mornet P, Orue J, Gillbert Y, Thiery G, Mamadou S (1956) La peste des petits Ruminants en Afrique occidentale française ses rapports avec la Peste Bovine. Revue d'elevage et de medecine veterinaire des pays tropicaux 9:313–342

Moustafa T (1993) Rinderpest and peste des petits ruminants-like disease in the Al-Ain region of the United Arab Emirates. Rev Sci Tech Off Int Epiz 12(3):857–863

Mulindwa B, Ruhweza SP, Ayebazibwe C, Mwiine FN, Muhanguzi D, Olaho-Mukani W (2011) Peste des petits ruminants serological survey in Karamoja sub region of Uganda by competitive ELISA. Vet World 4(4):149–152

Munir M, Abubakar M, Zohari S, Berg M (2012a) Serodiagnosis of peste des petits ruminants virus. In: Al-Moslih M (ed) Serological diagnosis of certain human, animal and plant diseases, vol 1. InTech, Croatia, pp 37–58

Munir M, Siddique M, Ali Q (2009) Comparative efficacy of standard AGID and precipitinogen inhibition test with monoclonal antibodies based competitive ELISA for the serology of peste des petits ruminants in sheep and goats. Trop Anim Health Prod 41(3):413–420

Munir M, Zohari S, Saeed A, Khan QM, Abubakar M, LeBlanc N, Berg M (2012b) Detection and phylogenetic analysis of peste des petits ruminants virus isolated from outbreaks in Punjab, Pakistan. Transboundary Emerg Dis 59(1):85–93

Munir M, Zohari S, Suluku R, Leblanc N, Kanu S, Sankoh FA, Berg M, Barrie ML, Stahl K (2012c) Genetic characterization of peste des petits ruminants virus, sierra leone. Emerg Infect Dis 18(1):193–195

Nyamweya M, Ounga T, Regassa G, Maloo S (2008) Technical brief on Pestes des Petits ruminants (PPR), ELMT Livestock Services Technical Working Group

Obidike R, Ezeibe M, Omeje J, Ugwuomarima K (2006) Incidence of peste des petits ruminants haemagglutinins in farm and market goats in Nsukka, Enugu state, Nigeria. Bull Anim Health Prod Afr 54:148–150

OIE (2011) Peste des petits ruminants, Tunisia. http://web.oie.int/wahis/public.php?page=single_report&pop=1&reportid=11864. Accessed 22 May 2010

OIE (2011a) Immediate notification report. Ref OIE: 10384, Report Date: 20/03/2011, Country: Algeria

OIE (2011b) Peste des petits ruminants, Tunisia

Pegram RG, Tereke F (1981) Observation on the health of Afar livestock. Ethiop Vet J 5:11–14

Perl S, Alexander A, Yakobson B, Nyska A, Harmelin A, Sheikhat N, Shimshony A, Davidson N, Abramson M, Rapoport E (1994) Peste des petits ruminants (PPR) of sheep in Israel: case report. Israel J Vet Med 49(2):59–62

Pervez K, Ashraf M, Khan MS, Khan MA, Hussain MM, Azim F (1993) A rinderpest like disease in goats in Punjab, Pakistan. Pak J Livest Res 1(1):1–4

Pest des Petits Ruminants in Iraq (2003) FAO Corporate document repository. http://www.fao.org/DOCREP/003/X7341E/X7341e01.htm. Accessed 13 March 2012

ProMED (2008) Peste des petits ruminants—Morocco (07): OIE, PPRV lineage IV. http://www.promedmail.org/pls/apex/wwv_flow.accept. Accessed 22 May 2010

Provost A, Maurice Y, Bourdin C (1972) La peste des petits ruminants: existe-t-elle en Africque centrale? Paper presented at the 40th general conference of the committee of the O.I.E.

Radostits OM, Gay CC, Blood DC, Hinchcliff KW (2000) Veterinary medicine, 9th edn. Elsevier, W. B. Saunders Co., London

Raghavendra AG, Gajendragad MR, Sengupta PP, Patil SS, Tiwari CB, Balumahendiran M, Sankri V, Prabhudas K (2008) Seroepidemiology of peste des petits ruminants in sheep and goats of Southern Peninsular India. Rev Sci Tech 27(3):861–867

Rahman MA, Shadmin I, Noor M, Parvin R, Chowdhury EH, Islam MR (2011) Peste des petits ruminants virus infection of goats in Bangladesh: pathological investigation, molecular detection and isolation of the virus. Bangladesh Vet 28(1):1–7

Rasheed IE (1992) Isolation of PPRV from Darfur state. University of Khartoum, Sudan

RO-CEA (2008) Horn of Arfica: Africa preparedness and response to the impact of soaring food prices and drought

Roeder PL, Abraham G, Kenfe G, Barrett T (1994) PPR in Ethiopian goats. Trop Anim Health Prod 26:69–73

Roger F, Guebre YM, Libeau G, Diallo A, Yigezu LM, Yilma T (2001a) Detection of antibodies of rinderpest and peste des petits ruminants viruses (Paramyxoviridae, Morbillivirus) during a new epizootic disease in Ethiopian camels (*Camelus dromedarius*). Rev Med Vet (Toulouse) 152:265–268

Roger F, Libeau G, Yigezu LM, Grillet C, Sechi LA, Mebratu GY, Diallo A (2000) International symposia on veterinary epidemiology and economics (ISVEE) proceedings, ISVEE 9. In: Proceedings of the 9th symposium of the international society for veterinary epidemiology and economics, Epidemiologic methods & theory session, Breckenridge, Colorado, USA, p 195

Roger F, Yesus MG, Libeau G, Diallo A, Yigezu LM, Yilma T (2001b) Detection of antibodies of rinderpest and peste des petits ruminants viruses (Paramyxoviridae, Morbillivirus) during a new epizootic disease in Ethiopian camels (*Camelus dromedarius*). Rev Med Vet (Toulouse) 152:265–268

Saeed IK, Ali YH, Khalafalla AI, Rahman-Mahasin EA (2010) Current situation of peste des petits ruminants (PPR) in the Sudan. Trop Anim Health Prod 42(1):89–93

Saha A, Lodh C, Chakraborty A (2005) Prevalence of PPR in goats. Indian Vet J 82:668–669

Santhosh A, Raveendra H, Isloor S, Gomes R, Rathnamma D, Byregowda S, Prabhudas K, Renikprasad C (2009) Seroprevalence of PPR in organised and unorganised sectors in Karnataka. Indian Vet J 86:659–660

Sanz-Alvarez J, Diallo A, De La Rocque S, Pinto J, Thevenet S, Lubroth J (2009) peste des petits ruminants (PPR) in Morocco. EMPRES Watch, 2008

Shaila MS, Purushothaman V, Bhavasar D, Venugopal K, Venkatesan RA (1989) Peste des petits ruminants of sheep in India. Vet Rec 125(24):602

Shaila MS, Shamaki D, Forsyth MA, Diallo A, Goatley L, Kitching RP, Barrett T (1996) Geographic distribution and epidemiology of peste des petits ruminants virus. Virus Res 43(2):149–153

Singh RP, Saravanan P, Sreenivasa BP, Singh RK, Bandyopadhyay SK (2004) Prevalence and distribution of peste des petits ruminants virus infection in small ruminants in India. Rev Sci Tech 23(3):807–819

Sow A, Ouattara L, Compaore Z, Doulkom B, Pare M, Poda G, Nyambre J (2008) Serological prevalence of peste des petits ruminants virus in Soum province, north of Burkina Faso. Revue d'elevage et de medecine veterinaire des pays tropicaux 1:5–9 (in French)

Sumption KJ, Aradom G, Libeau G, Wilsmore AJ (1998) Detection of peste des petits ruminants virus antigen in conjunctival smears of goats by indirect immunofluorescence. Vet Rec 142(16):421–424

Swai ES, Kapaga A, Kivaria F, Tinuga D, Joshua G, Sanka P (2009) Prevalence and distribution of peste des petits ruminants virus antibodies in various districts of Tanzania. Vet Res Commun 33(8):927–936

Tatar NK (1998) Ve keçilerde küçük ruminantlarin vebasi ve sigir vebasi enfeksiyonlarinin serolojik ve virolojik olarak arastirilmasi. Ankara Üniversitesi, Saglik Bilimleri Enstitusu, Ankara

Taylor WP, Al Busaidy S, Barrett T (1990) The epidemiology of peste des petits ruminants in the Sultanate of Oman. Vet Microbiol 22(4):341–352

Taylor WP, Barrett T (2010) Peste de Petits ruminants and rinderpest in: diseases of Sheep 4th edn. Blackwells Science, Oxford

Wambura PN (2000) Serological evidence of the absence of peste des petits ruminants in Tanzania. Vet Rec 146(16):473–474

Wamwayi HM, Rossiter PB, Kariuki DP, Wafula JS, Barrett T, Anderson J (1995) Peste des petits ruminants antibodies in east Africa. Vet Rec 136:199–200

Wang Z, Bao J, Wu X, Liu Y, Li L, Liu C, Suo L, Xie Z, Zhao W, Zhang W, Yang N, Li J, Wang S, Wang J (2009) Peste des petits ruminants virus in Tibet, China. Emerg Infect Dis 15(2):299–301

Waret-Szkuta A, Roger F, Chavernac D, Yigezu L, Libeau G, Pfeiffer DU, Guitian J (2008) Peste des petits ruminants (PPR) in Ethiopia: analysis of a national serological survey. BMC Vet Res 4:34

Whitney JC, Scott GR, Hill DH (1967) Preliminary observations on a stomatitis and enteritis of goats in southern Nigeria. Bull Epizoot Dis Afr 15:31–41

Yesilbag K, Yilmaz Z, Golcu E, Ozkul A (2005) Peste des petits ruminants outbreak in western Turkey. Vet Rec 157(9):260–261

Zahur AB, Irshad H, Hussain M, Ullah A, Jahangir M, Khan MQ, Farooq MS (2008) The epidemiology of peste des petits ruminants in Pakistan. Rev Sci Tech 27(3):877–884

Zeidan M (1994) Diagnosis and distribution of PPR in small ruminants in Khartoum state during 1992–1994. University of Khartoum, Sudan

# Chapter 6
# Current Advances in Molecular Diagnosis and Vaccines for Peste des Petits Ruminants

**Abstract** There has been a substantial improvement in the detection of the nucleic acids of Peste des Petits Ruminants Virus (PPRV), and effective vaccines have been developed in recent years. Demonstrations of several real-time PCR assays have provided powerful and novel means of not only detection but also quantification of the nucleic acids of PPRV in several types of clinical samples. Although most of the lineages are continent specific, reports on mixed lineages are emerging, such as in Sudan and Uganda. None of the available assays is devised so far to differentiate all of the lineages. Despite essential advances in the marker vaccines, it is still required to establish reverse genetics systems to rescue recombinant PPRV vaccines, primarily by manipulating the genes of PPRV and insertion of positive or negative markers, which ultimately will lead to the development of companion test. Similarly, swapping different genes will facilitate the establishment of a test that can differentiate vaccinated and infected animals. In this chapter, we discuss available diagnostic tests and potent PPRV vaccines. Furthermore, the current advances in both fields and possibilities to develop next-generation assays and vaccines for PPRV are critically discussed.

**Keywords** Diagnosis · Serology · Antigen detection · PCR · Vaccines · Multivalent vaccines · DIVA tests · Marker vaccines

## 6.1 Introduction

Peste des Petits Ruminants Virus (PPRV) is one of the most devastating respiratory diseases of small ruminants, which has spread in Sub-Saharan Africa, the Arabian Peninsula, and Southeast Asia. There are several assay formats available, such as ELISAs that can sensitively detect antibodies raised against PPRV.

M. Munir et al., *Molecular Biology and Pathogenesis of Peste des Petits Ruminants Virus*, SpringerBriefs in Animal Sciences, DOI: 10.1007/978-3-642-31451-3_6, © The Author(s) 2013

Similarly, polymerase chain reaction (PCR) has proved invaluable for the analysis of the PPRV genome even in poorly preserved field samples, and is relatively fast in order to timely diagnose the disease. Several assays are also available that can detect viral antigens in host tissues and secretions. A live attenuated tissue culture rinderpest (RP) vaccine has previously been used to prevent naïve populations from PPRV infection which is now restricted due to RP eradication campaign. With this, a homologous vaccine against Nigeria/75/1 and three Indian isolates has been devised and applied in field conditions. In order to differentiate vaccinated and naturally infected animals (DIVA), efforts have been made to establish marker vaccines. Due to the unavailability of a reverse genetic system for PPRV, most of the manipulations have been done in the infectious clone of the RP virus, which provided essential information not only regarding the potential of marker vaccines but also the abilities of the PPRV proteins to be used for immunizing small ruminants. In this chapter, all of the aspects, such as molecular bases for serological diagnosis, current advances in antigen and genome detection of PPRV are discussed. The quality of already available and recent improvements in PPRV vaccines is the primary focus of this chapter. Furthermore, future possibilities for the DIVA and multivalent vaccines are also be addressed.

## 6.2  Diagnosis of PPRV

### 6.2.1  Serological Diagnosis of PPRV

#### 6.2.1.1  Importance and Application

Early detection of PPRV is the best method by which veterinarians may attempt vaccination or other symptomatic treatment programs. In endemic countries, especially in rural areas where the best veterinary services are lacking, PPR is often misdiagnosed with many other bacterial, viral, and nutritional diseases with similar clinical pictures (see Sect. 3.3). This is primarily due to the lack of awareness about the disease, but also the unavailability of suitable diagnostic tools that are applicable and practicable in ordinary diagnostic laboratories. Effective implementation of control measures for PPR requires that diagnosis of the disease be made as quickly as possible, to contain outbreaks and minimize economic losses. Predominantly, the diagnosis of PPR in small ruminants is done serologically. Seropositivity is a good indication because animal infected with the PPRV carry antibodies for life, with the development of a sustained antibody response.

During the last few years, the detection of PPR antibodies by ELISA has been described and used in some countries, where this test is commercialized under various formats. However, ELISA kits may not be used as a mass-screening test due to its high cost, unless kits are developed and produced within the country. In this scenario, a reliable serological test that can be performed in ordinary diagnostic laboratories, as an alternative, is needed. Considering this as an objective, it is important to estimate the performance characteristics (Kappa values, relative diagnostic sensitivity, and specificity, etc.) of tests that are intended to replace the expensive tests.

Serological tests are often the method of choice for mass screening of populations, with their main limitation being the failure to demonstrate antibodies, i.e., sensitivity. While nucleic acids amplification methods, such as polymerase chain reaction in conventional and real-time formats offer greater sensitivity, they are usually too expensive for routine diagnosis in many laboratories, particularly in developing countries (Muthuchelvan et al. 2006). Such is also the case with in vitro isolation in cell culture, owing to high operating costs, quality assurance issues and lack of trained scientists and suitable facilities. Rapid assays, such as immunochromatographic or magnetic bead format for the detection of antigens or antibodies, which are simple to perform and interpret and can be performed onsite or close to the ''farm'', offer more practical solutions in the developing world. However, such rapid tests for the diagnosis of PPR are not currently available. The development of such technologies using the precipitation line on gel principles in a rapid, cheap, and accurate format is of great assistance to disease control authorities in many developing countries. Agar gel immunodiffusion test (AGID), which is recommended by the Office International des Epizooties (OIE) for antigen detection, is currently in use in many countries for the identification of PPRV. This test is relatively rapid, inexpensive, and simple but is not highly sensitive to identify animals infected with the PPR virus, although it is interpreted subjectively by visual reading of a precipitation line curvature.

### 6.2.1.2 Bases for Serological Diagnosis

Most of the serological assays are based on the detection of antibodies against nucleocapsid (N) and hemagglutinin-neuraminidase (HN) proteins of PPRV.

### N Protein

In most of the single stranded RNA viruses, including PPRV, the N protein is highly conserved and the most immunogenic protein. Being close to the $3'$ end of the genome of PPRV, it is produced in quantities higher than any other structural proteins of morbilliviruses, due to attenuation that occurs at each intergenic region between two genes (Lefevre et al. 1991) (Sect. 1.2.2). The antibodies produced against the N protein do not protect the animals from the disease, but being most immunogenic and abundant, it remains the most acceptable target for the design of PPRV diagnostic tools (Diallo et al. 1994). Additionally, the N protein of PPRV appears to be both type specific and have cross-reactive epitopes. The N protein of PPRV has been divided into four regions: region I (aa 1–120), region II (aa 122–145), region III (aa 146–398), and region IV (421-445). The most immunogenic epitopes have been mapped in region I and II, whereas region III and IV are the least immunogenic (Choi et al. 2005). Another study noted that, the amino acids from 452 to 472 are the most immunogenic part within the N protein. It has further been summarized that there is a development of an earlier immune response to region I and II than to region III and IV (Bodjo et al. 2007). A recombinant baculovirus that expresses the N protein in insect cells or larvae (*Spodoptera frugiperda*) (Ismail et al. 1995), or in *Escherichia coli* (Yadav et al. 2009) has been used successfully as a coating antigen in an enzyme-linked immunosorbent assay for the serological diagnosis of PPRV. On the other hand, a cell culture attenuated live PPRV is used as antigen in both competitive ELISA (cELISA) (Singh et al. 2004b) and sandwich ELISA (sELISA) (Singh et al. 2004a). Taken together, most of the diagnostic assays for PPRV have been developed based on the monoclonal antibodies (mAb) raised against the N protein (Libeau et al. 1995).

HN Protein

The HN protein of PPRV is the most diverse among all the members of morbilliviruses. Comparison among morbilliviruses have indicated that RPV and PPRV, the two most similar members of the genus morbilliviruses, share only 50 % similarity in their HN proteins. The most variable nature of HN protein probably reflects the role of this protein in species specificity. If this is the case, then the H proteins of rinderpest virus and PPRV may have significant potential in DIVA strategies (differentiation of infected from vaccinated animals). Since the HN protein determines cell tropism, most of the protective host immune response is raised against HN protein. Contrary to the N protein, it has been investigated that antibodies raised against recombinant HN protein are protective enough to prevent disease in case of PPRV infection. Alternatively, it is possible to establish a DIVA strategy by immunization of a naïve population with recombinant protein expressing only HN protein, were an ELISA against the N protein of PPRV will act as a DIVA test. For these reasons and an attraction to the neutralizing antibodies against HN protein, it has remained under continuous immunological pressure. The HN protein is not only involved in cell tropism, but studies indicate that it may have a role as a neuraminidase. PPRV is unique among morbilliviruses, which carry this function. Mapping of the functional domain, using monoclonal antibodies, have demonstrated that two regions, one at amino acid 263–368 and other at 539–609 amino acids, are the most immunodominant epitopes (Seth and Shaila 2001). There is an increasing tendency to design DIVA strategies targeting the HN protein of PPRV.

## 6.2.2 Serological Assays

Based on the above-mentioned facts, ELISAs have been developed targeting the HN (Anderson and McKay 1994; Saliki et al. 1993; Singh et al. 2004b) and N proteins (Libeau et al. 1995) for specific detection of antibodies against PPRV, both in ovine and caprine hosts. The ELISA using the N protein antigen is based on the competition between the tested sample antibodies and MAb against a specific epitope on the N protein. A level of 45 % competition was observed in negative controls due to lack of cross-reaction with the N-protein (Libeau et al. 1995). A virus neutralization test (VNT) was found to be more sensitive than the N antigen-based ELISA, which is probably due to lack of virus neutralization abilities of the antibodies produced against the N protein (Diallo et al. 1995). The relative sensitivity and specificity between the VNT and cELISA were found to be 94–95 and 99.4 %, respectively. The ELISAs using the HN protein of PPRV have been applied for the development of both blocking and competitive ELISAs. Although both these ELISAs are based on the competition between serum antibodies and MAbs against the HN protein of PPRV, the test sera are preincubated with antigen followed by incubation with MAbs (Saliki et al. 1993). The sensitivity and specificity have been estimated to be 90.4 and 98.9 %, respectively. The detailed principle, experimental protocol, and molecular bases have been recently reviewed comprehensively (Munir 2011; Munir et al. 2012a).

VNT is a highly sensitive and specific test, and is probably the most reliable test to demonstrate antibodies against any member of the genus morbillivirus in test sera. However, it is time consuming, expensive, and labor intensive. VNT is usually performed in primary cell lines (e.g. lamb kidney cells) grown in roller-tube cultures. It is noted, due to cross-neutralization abilities of PPRV and RPV, that serum samples from RPV-infected animals may neutralize the PPRV in this test. However, the level of neutralization was observed to be higher in homologous viruses (antibodies against PPRV neutralize PPRV) than heterologous viruses (antibodies against PPRV neutralize RPV). Therefore, reciprocal cross neutralization may be applied to differentiate PPRV from RPV-infected animals (Taylor and Abegunde 1979). There have been several alternative tests reported for the serological diagnosis of PPRV, such as indirect N ELISA (Ismail et al. 1995), immunofiltration (Dhinakar Raj et al. 2000), sandwich ELISA (Saravanan et al. 2008), hemagglutination tests (Dhinakar Raj et al. 2000; Manoharana et al. 2005), latex agglutination tests (Keerti et al. 2009), single radial hemolysis test (Munir et al. 2009a), and a precipitation inhibition test (b). The application of all these assays in terms of their sensitivity and specificity was recently compared and critically reviewed (Munir 2011; Munir et al. 2012a).

## 6.2.3  Antigen Detection for PPR Diagnosis

Two different formats of ELISA have been developed and applied in the field to efficiently detect antigens in the tissues and secretions of PPRV-infected animals. Immunocapture ELISA (Libeau et al. 1994) overwhelmed sandwich ELISA (Saliki et al. 1994), even until now over, however, both utilized MAbs directed against the N protein of PPRV. Both assays are rapid (performed within 2 h), sensitive, and specific (detection level of $10^{0.6}$ $TCID_{50}$/well), simple (in the format of precoated plate), and robust (able to detect antigens in the samples not kept under ideal conditions). Because the MAbs used in these assays are raised against the overlapping and common domains of the N protein of PPR and RP viruses, this assay can be used to differentiate PPRV from RPV-infected animals (Libeau et al. 1994).

The AGID and counter immunoelectrophoresis (CIEP) have consistently been practiced for the detection of both antigen and antibodies, in <4 h (Obi 1984). These assays are reliable, simple, and fast to screen various biological samples for PPRV, and may serve as alternatives to expensive and labor intensive assays (Munir 2011; Munir et al. 2012a; Munir et al. 2009b). Immunofluorescence (Sumption et al. 1998) and immunochemistry (Eligulashvili et al. 2002) have also been applied successfully for the demonstration of PPRV antigens, and are compared in parallel to antigen detection ELISAs (Munir 2011).

## 6.2.4  Genome Detection for PPR Diagnosis

Although assays to detect antibodies or antigens are promising, tests such as VNT or virus isolation require biologically active material, and assays such as ELISA require animal sera in relatively well-preserved format. PCR has been proved invaluable for the analysis of poorly preserved field samples, and is relatively fast in order to timely diagnose the disease. Since the genome of all members of the family paramyxoviridae is single-stranded RNA, it is essential to reverse transcribe into complementary DNA (cDNA). Initially, Forsyth and Barrett (1995) demonstrated the reverse transcription (RT) of the PPRV genome followed by PCR targeting the F protein mRNA (Table 6.1). At this time, due to limitations of sequence availability for other genes, and because F gene was lavishly used for phylogenetic analysis, it was considered that the F gene might be the best target (Forsyth and Barrett 1995). However, they observed inconsistency in the performance of different sets of primers designed for both the F and P genes. They concluded that the assay is not suitable for every virus strain, variant or isolate, due to changes at the $3'$ end of the primer binding sites, which may yield a false-negative result. Moreover, RNA viruses are subject to high mutation (nucleotide substitution) error frequencies (Steinhauer et al. 1989), and therefore are likely to escape detection. It is, thus, crucial to target those genes that are highly conserved in different strains of PPRV. Nevertheless, they have provided substantial information for the application of RT-PCR, not only for differentiation of PPRV from RPV but to characterize them for the phylogenetic analysis. They have further proposed that the N or L gene of PPRV may be better suited for detection of the PPRV genome, due to being more abundant than any other viral genes and being conserved among morbilliviruses. This was latter accomplished by Couacy-Hymann et al. (2002) who successfully amplified the 3′ end of the PPRV N gene mRNA and found this assay more sensitive than traditionally used Vero cell titration assays (Couacy-Hymann et al. 2002) (Table 6.1).

To avoid false-negative results in F gene-based RT-PCR, Balamurugan et al. (2006) presented a one-step single-tube multiplex RT-PCR targeting the N and M genes. Comparison of the sequences revealed that the matrix (M) proteins of PPRV and other morbilliviruses have a high degree of conservation for this protein sequence (Haffar et al. 1999). The M protein is reported to be the most conserved among all morbillivirus proteins (Sharma et al. 1992) (see Chap. 1, Table 6.1), and it is also synthesized most abundantly in the infected cells (Diallo 1990). There-fore, for the detection of the virus a PCR method based on viral genes that are close to the 3′ end of the PPRV genome may be more appropriate than a PCR based on the F protein gene, which is further from 3′ end than the M gene of PPRV. The primers were designed so that the PPRV positive samples yield both N and M gene products, whereas only the N gene product (337 bp) will be seen when RPV is present. Based upon comparison with an sELISA, it was concluded that this RT-PCR is efficient in amplification of the PPRV N and M gene regions, for rapid detection and differentiation of PPRV from RPV in clinical samples, with increased sensitivity and reduced false positivity (Balamurugan et al. 2006).

The above-described PCRs carry some general limitations: these are labor intensive, require visualization of the PCR products on gels, there is a high risk of contamination, and they are not suitable for high-throughput testing. It was not until 2008 when Bao and co-workers developed a very sensitive and specific TaqMan based, one-step real-time quantitative reverse transcription PCR (qRT-PCR) for the detection of PPRV in field samples (Bao et al. 2008), which proved to be very useful for the analysis of samples collected during PPR epidemic in Tibet in 2007. This assay overwhelmed the existing RT-PCRs for rapid, specific, and sensitive laboratory detection of PPRV in tissue samples from field cases (Bao et al. 2008) (Table 6.1). However, the test was not validated on all PPRV lineages, and its performance was not established clearly on field samples. Therefore, with an aim to detect and quantify all four lineages of PPRV in field samples, Kwiatek et al. (2010) designed primers and probes in the 3′-end variable nucleotide sequence of the morbillivirus N gene, which has been used to phylogenetically define the lineages of PPRV. They further concluded that this assay provides sensitive and specific detection of all PPRV lineages, including those currently circulating in Africa, the Middle East, and Asia. Additionally, it is quicker, thus allowing high speed/high-throughput monitoring of susceptible small ruminants (Kwiatek et al. 2010) (Table 6.1).

None of the above-mentioned PCR is a field-based assay, primarily due to the need for thermocycler and electrophoresis apparatus for RT-PCR, and expensive

**Table 6.1** The properties and comparison of different formats of PCR routinely used for the detection of PPRV

| Name of the primers and probes | Format of the assay | Forward (F) Reverse (R) Probe (P) (5′-3′ end) | Target genes: Fusion (F) Nucleocapsid (N), Matrix (M) | Location in the corresponding gene | Product size (bp) | Detection limit | Comparison with other technique | Reference |
|---|---|---|---|---|---|---|---|---|
| F1 | RT-PCR | **F** = ATCACAGTGTTAAA-GCCTGTAGAGG | F | 777–801 | 371 | RT-PCR (12/23)[a] AGID and Virus isolation (0/23) | AGID and virus isolation | Forsyth and Barrett, (1995) |
| F2 | | **R** = GAGACTGAGTTTGTG-ACCTACAAGC | | 1,124–1,148 | | | | |
| NP3 | RT-PCR | **F** = TCTCGAAATCGCCTC-ACAGACTG | N | 1,232–1,255 | 350 | $10^{-3}$ $TCID_{50}$/ml | Vero cell titration test | Couacy-Haymann et al. (2002) |
| NP4 | | **R** = CCTCCTCCTGGTCCTC-CAGAATCT | | 1,583–1,560 | | | | |
| Fr2 | Multiplex RT-PCR | **F** = ACAGGCGCAGGTTT-CATTCTT | N | 1,270–1,290 | 337 | Multiplex RT-PCR (22/32) | Sandwich ELISA | Balamurugan et al. (2006) |
| Re1 | | **R** = GCTGAGGATATCCTT-GTCGTTGTA | M | 1,584–1,606 477–497 | 191 | | | |
| MF-Morb | | **F** = CTTGATACTCCCCAG-AGATTC | | 646–667 | | | | |
| MR PPR 3 | | **R** = TTCTCCCATGAGCCG-ACTATGT | | | | Sandwich ELISA 18/32 | | |
| PPRNF | Real-time PCR | **F** = CACAGCAGAGGAAGC-CAAACT | N | 1,213–1,233 | 94 | Detection range from $8.1$–$8.1 \times 10^9$ RNA copies. It is sensitive with one log unit than conventional RT-PCR | Conventional RT-PCR (Couacy-Haymann et al. (2002)) | Bao et al. 2008 |
| PPRNP (Probe) | | **P** = FAM-5-CTCGGAA-ATCGCC TCGCAGGCT-5-TAMRA | | 1,237–1,258 | | | | |
| PPRNR | | **R** = TGTTTGTGCTGGAG-GAAGGA | | 1,327–1,307 | | | | |

(continued)

**Table 6.1** (continued)

| Name of the primers and probes | Format of the assay | Forward (F) Reverse (R) Probe (P) (5'-3' end) | Target genes: Fusion (F) Nucleocapsid (N), Matrix (M) | Location in the corresponding gene | Product size (bp) | Detection limit | Comparison with other technique | Reference |
|---|---|---|---|---|---|---|---|---|
| NPPRf | Real-time PCR | F = GAGTCTAGTCAAAAACC-CTCGTGAG | N | 1,438–1,461 | 96 | 32 copies per reaction with a corresponding Ct value of 39 | Conventional RT-PCR (Couacy-Haymann et al. (2002)) | Kwiatek et al. (2010) |
| NPPRp (probe) | | P = FAM-5-CGGCTGAGGCACTCTT-CAGGCTGC-3'-BHQ1 | | 1,472–1,495 | | | Real-Time PCR (Bao et al. (2008)) | |
| NPPRr | | **R** = TCTCCCTCCTCCTGGTCCTC | | 1,516–1,534 | | | | |
| F3 | LAMP | TTGCAATGCAGTCAACCT | M | 420–437 | 217 | $1.41 \times 10^{-4}$ ng total RNA per assay. 10-fold higher than that for RT-PCR, and similar to that of the real-time RT-PCR assay | Real-Time PCR (Bao et al. (2008)) | Li et al. (2010) |
| B3 | | ATTCTCCCATGAGCCGA | | 620–636 | | | | |
| F1c | | GCACACTATAGTAACCATTGTCTGA | | 496–520 | | | | |
| F2 | | TGATACTCCCCAGAGGTT | | 447–464 | | | | |
| B1c | | GGAGTTCCGCTCAGCCAATG | | 534–553 | | | | |
| B2 | | TTCTAGGGTTTGTGCCATT | | 592–610 | | | | |
| LF | | TCTAGTTATGCTCATGTACACAACC | | 468–492 | | | | |
| LB | | GTAGCCTTCAACATCTTGGTTACAC | | 556–580 | | | | |

[a] 12 samples out of 23 were detected positive with RT-PCR

real-time PCR for probe-based assays. To counteract this problem, loop-mediated isothermal amplification (LAMP) was proposed to be a suitable alternative. LAMP is based on the principles of a strand displacement reaction, and the stem-loop structure amplifies the target with high specificity, selectivity, and rapidity under isothermal conditions (Nagamine et al. 2002), which provides a fast and sensitive method to amplify virus RNA, obviating the need for a thermal cycler. The higher amplification efficiency of the RT-LAMP method enables simple visual observation of amplification with the naked eye in the presence of an intercalating dye, such as SYBR Green I or ethidium bromide. This assay approved to be highly sensitive for the detection of PPRV from all the continents (Li et al. 2010; Wei et al. 2009) (Table 6.1).

For onsite application, it is highly plausible to combine the simple procedures for RNA template preparation, such as a Whatman FTA card and FTA purification reagent (Munir et al. 2012b; Munir et al. 2012c), with the RT-LAMP assay, which could easily be applied for field diagnosis of PPRV. Despite the high sensitivity and specificity of these assays, and their validity to detect both vaccine and field viruses, none of the assays is a formally approved OIE method. Therefore, they require extensive validation before approval.

## 6.3 Vaccines Against PPRV

Due to the nature of replication in host lymph nodes and the possibility to disarm the host defence mechanisms, PPRV along with many other morbilliviruses are profoundly but transiently immunosuppressive. This immunosuppression is characterized by concurrent infections, and subsequently leads to high mortality. Despite significant immunosuppression, recovery from the infection is usually followed by the establishment of a strong, specific, and long-term protective immune response by the host (Cosby 2005). The characterization of protective immunity against measles virus (MV), RPV and canine distemper virus (CDV) is well described. However, in the case of PPR information is lacking with regard to the immune response necessary for recovery from or for protection against infection (see Chap. 4). Due to the high functional and structural similarities between morbilliviruses, vaccines against PPRV were developed or are being developed following the same strategy as for the other morbilliviruses. There have been significant improvements in vaccine developments to control PPRV, which can conveniently be divided into four categories.

### 6.3.1 Heterologous Attenuated PPRV Vaccines

When PPRV was first identified, it was speculated that PPRV is a variant of RPV, for several reasons such as that PPRV vaccine can protect cattle against disease caused by PPRV, and rinderpest antiserum can reduce the titer of PPRV in neutralization tests. However, due to the absence of cross-neutralization it was latter proposed that PPRV is distinct from RPV (Hamdy et al. 1976); but a study conducted by Taylor and Abegunde concluded that both viruses cross neutralize each other (Taylor and Abegunde 1979). Finally, Gibbs et al. (1979), based on cross protection and cross-neutralization among several morbilliviruses, suggested that PPRV has its own identity (Gibbs et al. 1979). Subsequent studies revealed, based on monoclonal antibodies, that PPRV is closely related to RPV; however, later gene sequence analysis indicated that RPV and MV are more closely related than RPV and PPRV (McCullough et al. 1986). It is also believed that RPV is an archetype virus from which most of the currently identified morbilliviruses are evolved (Norrby et al. 1985).

Due to a failure to develop an attenuated PPRV for immunization after 65 cell-culture passages (Gilbert and Monnier 1962), Bourdin et al. (1970) and Bonniwell (1980) successfully conducted a field trial using RP vaccine to protect animals against PPRV (Bourdin et al. 1970; Bonniwell 1980). Ultimately, due to confirmed cross-protection between RPV and PPRV, and the availability of RP vaccines at the time of first recognition of PPRV, an attenuated Plowright's tissue culture RP vaccine (TCRPV) was shown to protect animals from PPRV in many countries. This vaccine was considered safe in pregnant goats (Adu and Nawathe 1981), and the upcoming kids carried passive immunity for at least 3 months. The vaccinated animals are protective against PRPV for at least 3 years (Rossiter 2004), which is a consequence of strong cross-cellular immune responses. However, due to intense planning of RPV eradication and to attain the status of RP free countries, the use of such vaccine was discontinued.

## 6.3.2  Homologous Attenuated PPRV Vaccines

With the restriction of TCRPV vaccines, there were immense efforts started to develop homologous vaccines against PPRV. After a series of failures, Gilbert and Monnier (1962) for the first time, successfully grew PPRV in primary cell culture, where they observed the large syncytia formation as a cytopatopathic effect (CPE) (Gilbert and Monnier 1962). Later, some other CPEs, such as refringent and rounded cells, were manifested as CPE specific to PPRV (Laurent 1968). Using hematoxylin and eosin staining, mini- and micro-syncytia formations were visualized as makers for PPRV replication, especially at the initial stages of infection, as has previously been observed for RP (Plowright and Ferris 1959). However, despite early virus isolation, the attenuation was not possible even after 65 passages, until Diallo et al. (1989) reported a PPRV that is attenuated in cell culture, and established the bases for a homologous vaccine against PPRV (Diallo et al. 1989). The PPRV isolated by Taylor and Abegunde (1979) from Nigerian goats which had died from PPRV infection in 1975, was adapted to Vero cells at 37 °C and later, even until now, approved as a prototype for PPRV (Taylor and Abegunde 1979).

In the process of PPRV adaptation in cell culture (Vero cells), the PPRV passaged for 20 times show mild clinical disease characterized by hyperthermia when inoculated in susceptible population (Taylor and Abegunde 1979; Diallo et al. 1989). However, after 35 more passages (55 passages in total), the pathogenicity totally diminished, which lead to the development of an avirulent strain suitable for immunization. Goats infected with this virus not only remained healthy but also survived a challenged virus, and the virulence was irreversible until three consecutive back passages in live animals. The virus remained avirulent until 120 passages in Vero cells, and the inoculated animals were unable to transmit the virus to healthy unvaccinated animals. In the next step of a vaccine trial in the field, an extensive application of this vaccine (at the 63rd passage) was practised from 1989 to 1996. The combined results, based on 98,000 sheep and goats of which 58,000 were vaccinated, indicated that approximately 98 % of the vaccinated animals seroconverted after 1 month and remained protective for at least 3 years. The effective dose was calculated to be $10^{0.8}$ TCID50/animal; however, a dose of $10^3$ TCID50/animal also proved to be safe (Martrenchar et al. 1997). Pregnant animals remained safe and were able to pass passive immunity to their offspring, which remained protected for 3–5 months. Later, the use of vaccine in the field was shown to be protective against wild-type PPRV virus, and immunized animals are protective against RPV, despite low sensitivity to RPV infection.

The second successful attenuated PPRV virus is Sungri/96, which was isolated from goats that died with PPRV in the Sungri area in Himachal Pradesh, India during 1994 (Sreenivasa et al. 2002). This isolate was initially passaged for 10 times in B95a (marmoset lymphoblasoid) cells, followed by 49 passages in interferon deficient Vero cells. However, this isolate was attenuated completely after 56 passages directly in Vero cells (Sarkar et al. 2003). Two other attenuated PPRV isolates, Arasur/87, and Coimbatore/97, were isolated from sheep and goats, respectively, and were attenuated after 75 passages in Vero cells (Saravanan et al. 2010).

Currently, a study conducted by Saravanan et al. (2010) made a comprehensive comparison of all these three Indian vaccines (Saravanan et al. 2010). They determined the sterility, safety and potency, and the post-vaccination immune status of the animals. The results demonstrated that the vaccines appeared to be sterile and safe at 100, 1 and 0.1 filed doses, and no untoward reactions were observed. All of the animals vaccinated with Sungri 1996 and Arasur 1987 vaccines withstood the challenge up to 14 days post challenge, without showing either rise in rectal temperature or other clinical signs specific to PPR. Additionally, the vaccines showed cross-specific protection with respect to sheep and goats. The swabs collected from these animals were negative for PPRV antigen, indicating that the vaccine was 100 % potent and efficacious both in sheep and goats. The data gathered on molecular characterization, immunosuppression (Rajak et al. 2005), thermostability (Sarkar et al. 2003), and safety (Saravanan et al. 2010) indicate that a single vaccination is sufficient to provide lifelong immunity in sheep and goats. It has also been suggested that the presently available vaccines could be exploited for mass vaccination of sheep and goats in developing countries.

## 6.3.3 Stability of Live Attenuated PPRV Vaccines

The main drawback to the above-mentioned vaccines remains thermostability, especially in the scenario when disease is only endemic in tropical countries. PPRV, as with other morbilliviruses, is heat labile, and therefore heat sensitivity poses a serious problem in the live attenuated vaccines in hot climate condition. In addition, since the disease is prevalent in most developing countries, it is difficult to maintain the cold chain to ensure the vaccine potency, due to poor infrastructure. All of these factors inevitably result in the loss of vaccine potency at the end time of its use in animals. To alleviate this drawback, it is needed to generate a heat-tolerant product. There have been several efforts to avoid this problem, either through construction of recombinant vaccines or to increase the longevity of the already existing live attenuated vaccines.

Lyophilization appeared to be a prevailing approach to stabilize the heat sensitive vaccines, especially in the presence of suitable excipients. Studies on the addition of lactalbumin hydrolysate-sucrose (LS), Weybridge medium (WBM), and lactalbumin hydrolysate-manitol (LM) with lyophilized PPRV strain Nigeria 75/1 vaccine have shown that the WBM formulation could maintain the virus titer for a longer time compared to live attenuated vaccine (Asim et al. 2008). Different stabilizers, i.e. LS, WBM, buffered gelatin-sorbitol (BUGS), and trehalose dihydrate (TD), were also used to prepare the PPRV Sungri 96 vaccine. The combined results showed that LS and TD allowed for higher stability of the lyophilized PPRV vaccine without compromising the safety of the vaccine (Sarkar et al. 2003). The LS stabilizer could also maintain the protective titer of the Vero cell adopted RP vaccine up to 4 h at room temperature, if reconstituted with 0.85 % sodium chloride and 1 M magnesium sulphate (Mariner et al. 1990). Moreover, an approach to stabilize the PPRV Nigeria 75/1 vaccine was the use of the dehydration method Xerovac in the presence of a formulation containing trehalose (Worrall et al. 2000). Under these conditions, the vaccine is stable at 45 °C for 14 days with minimal loss of potency. The World Organization for Animal Health (OIE) recommends the use of WBM as a stabilizing solution for the lyophilized PPRV vaccine. However, this vaccine formulation is still very susceptible to thermal degradation (Sarkar et al. 2003). The thermostability of live attenuated vaccines can be enhanced by the use of a suitable combination of stabilizers and heavy water, such as in polio and yellow fever vaccines (Wu et al. 1995; Adebayo et al. 1998). Additionally, the application of deuterium for enhancing the thermostability of PPRV increased when using heavy water as the reconstituting diluent. The use of heavy water-$MgCl_2$ as the reconstituting diluent in the PPRV vaccine increased the stability of $10^{2.5}$ TCID50/ml from 14 days in conventional PPRV vaccine to 28 days at 37 and 40 °C. In conclusion, deuterated virus reconstituted in heavy water-based diluent shows higher titers than conventional virus (Sen et al. 2010).

Besides the stability during vaccines storage and transportation, there are several factors that negatively influence the final virus potency during the vaccine production process. Being an enveloped virus, the stability of the live attenuated vaccine can be compromised in cell culture bulks due to temperature. It was reported that the intrinsic stability of PPRV live attenuated vaccines can be increased with the high concentration of glucose or fructose (Silva et al. 2008); and higher WBM osmolalities actually used for the production of this vaccine (Diallo 2004). In order to determine the role of sucrose and trehalose in WBM, Silva et al. (2011) have recently noted that in the presence of ris/trehalose liquid formulation the virus's half-life remained for 21 h and 1 month at 37 and 4 °C, respectively. However, in the lyophilized form, the same formulation was able to maintain the viral titer above the $1 \times 10^4$ TCID50/mL (>10 doses/mL) for at least 21 months at 4 °C (0.6 log lost), 144 h at 37 °C (0.6 log lost), and 120 h at 45 °C (1 log lost). The addition of 25 mM fructose resulted in a higher virus production (1 log increase) with higher stability (2.6-fold increase compared to glucose 25 mM) at 37 °C. Increased concentrations of NaCl improved virus release, reducing the cell-associated fraction of the virus produced. Moreover, this harvesting strategy is scalable and more suitable for a larger scale production than the freeze/thaw cycles normally used (Silva et al. 2011).

## 6.4 Recombinant Marker PPRV Vaccines

While the thermostability of the attenuated PPRV vaccines of either lineage I (Nigeria/75/1) or lineage IV (Sungri/96, Arasur/87 and Coimbatore/97) is being improved, it is important to design recombinant DNA vaccines. These maker vaccines are essential to maintain thermostability, to accommodate multivalent vaccines to protect several diseases, to differentiate infected from vaccinated animals (DIVA), and to provide efficient seromonitoring. Several studies have been conducted to build recombinant vaccines, swapping either antigenically related RPV genes (heterologous) or using the genes of PPRV (homologous), and their protective capabilities for PPRV have been compared, with variable results.

### 6.4.1 Heterologous Marker Vaccines for PPRV

By virtue of antigenic relatedness, the surface proteins (F and H/HN) of RPV and PPRV can be used interchangeably to protect animals for any of these viruses. In the time of RP as the main disease of livestock, a marker vaccine was constructed using a recombinant vaccinia virus. This construct contains the H and F genes of RPV, and was not only protective against RPV but also completely protective against PPRV, although neutralizing antibodies were only detected for RPV (Jones et al. 1993). These results indicate that the H and F proteins of RPV are sufficient for cross-protection for PPRV; however, it remained to be determined which protein immune response overwhelms the protection over the other. In this context, construction of a recombinant capripox virus containing either the H or F genes of RPV established complete protection against challenged PPRV (Romero et al. 1995). However, the efficacy of this vaccine against capripox, along with RPV or PPRV, was not evaluated. These constructs, indeed, give protection against PPRV, but they do not restrict the replication of PPRV especially at the beginning of immunization, probably due to partial immunity at earlier time points. Nevertheless, these remained successful heterologous marker vaccines, and can be used alternatively in case of continuous failure of efficient recombinant vaccines specifically designed for PPRV.

## 6.4.2 Homologous Marker Vaccines for PPRV

It is important to briefly mention that the HN protein of PPRV mediates attachment of the virus to the host cell membrane, whereas the F protein facilitates the virus entry (see Chap. 2). This process is essential for the spread of virus from one cell to other. Additionally, the F protein is critical for the induction of an effective and protective immune response. With this in mind, it is plausible to hypothesize that immunity against both of these proteins is essential to prevent the initiation of infection, and to abolish the dissemination of infection. Studies have been conducted to evaluate the abilities of individual proteins against challenge with PPRV, with somewhat controversial results.

Expression of F proteins of most of the morbilliviruses in either poxvirus or vaccinia virus by using recombinant DNA technology has proven to be effective as vaccines. Considering this property common for PPRV, a capripox virus recombinant that expresses the PPR F protein can protect goats against two diseases, PPR and capripox, at the same time when challenged with PPRV (subcutaneous injection of $10^4$ Guinea-Bissau/89) and capripox (intradermal inoculation of 0.2 ml of Yemen isolate). It was concluded that a low dose [(0.1 plaque forming unit (PFU)] requirement and protection for two diseases would reduce the cost of controlling these diseases by vaccination (Berhe et al. 2003).

Recently, it has been demonstrated that a recombinant extracellular baculovirus expressing the HN protein of PPRV generates virus neutralizing antibody responses, bovine leukocyte antigen (BoLA) class II restricted helper T cell responses and BoLA class I restricted cytotoxic T cell (CTL) responses. Moreover, the animals that were immunized with this construct were protective against PPRV, and the generated antibodies were neutralizing for both PPRV and RPV (Sinnathamby et al. 2004). It was concluded that the HN glycoprotein of PPRV is sufficient to mount long lasting humoral and cell-mediated immune responses in cattle of different breeds and parentage, and can hence serve as a potential subunit vaccine against RP as well as PPR. This immune dominant character was mapped to a highly homologous domain (amino acids 400–423) on HN protein of PPRV. Recent demonstration that baculoviruses can infect a wide variety of mammalian cells, without being able to replicate in such cells, may be beneficial for the efficient delivery of recombinant baculovirus-based vaccines to antigen presenting cells for better immune responses (Ghosh et al. 2002).

Primarily, focus has remained on the HN protein for the immunogenic properties of PPRV, whereas the immunogenic properties of the F protein have not been studied in detail. However, the F protein is of equal importance not only because, it plays a crucial role in viral infectivity but it is also the main player in protective immunity. This can be realized by the fact that hyperimmune serum induced against the F protein has the ability to inhibit the PPRV-induced cell fusion and F protein-mediated hemolysis (Devireddy et al. 1999). Moreover, expression of the F protein induces protective immunity against lapinized RPV (Devireddy et al. 1998). It is to note that the F protein of the paramyxoviruses has high amino-acid homology, predicting conserved biological activities (Lamb 1993).

To understand the immunogenic properties of the F protein and its potential as a marker vaccine candidate, Rahman et al. (2003) have constructed a recombinant *Bombyx mori* nucleopolyhedrovirus (BmNPV), which was able to express antigenic epitopes of the F protein of PPRV and the H protein of RPV (Rahman et al. 2003). It has been demonstrated that BmNPV expressing the F protein of PPRV or the H protein of RPV were displayed on the virion surface, as well as the surface of the virus-infected cells and adipose tissues of host larvae (Fig. 6.1). The immunogenicity of the H or F displayed on the recombinant BmNPV was also examined in adult male BALB/c mice. Upon intraperitoneal immunization with the purified recombinant viruses, high antibody titers were achieved against both proteins. Interestingly, the antibodies raised against each of the viral proteins (F and H/HN) were equal in neutralizing viruses of other species (Table 6.2). However, the virus neutralization titers of the displayed RPV-H antibodies were lower than that of PPRV-F, and it was also less effective against PPRV. Since PPRV is known to possess hemagglutination activity (Ramachandran et al. 1995), both PPRV-F and RPV-H antisera raised against the BmNPV-displayed antigens showed inhibition of hemagglutination activity of PPRV. The hemagglutination inhibition titer of PPRV-F antibodies was four times higher than that against RPV-H. There was no inhibition by preimmune serum or serum from mice infected with wild-type BmNPV, as expected (Table 6.2) (Rahman et al. 2003).

Based on these and other results, the authors were able to conclude that since host larvae infectivity was retained when infected with BmNPV expressing the F and H proteins of PPRV and RPV; respectively, it is possible to use the *B. mori* larvae for large-scale production of recombinant antigens in lieu of the cell-culture system. On the other hand, expression of the protein through the baculovirus display system allows rapid generation of effective antigens without the need for purifying the recombinant proteins, and the recombinant baculoviruses also have a good biosafety profile, and therefore vaccination using recombinant baculoviruses should be a safe approach (Rahman et al. 2003).

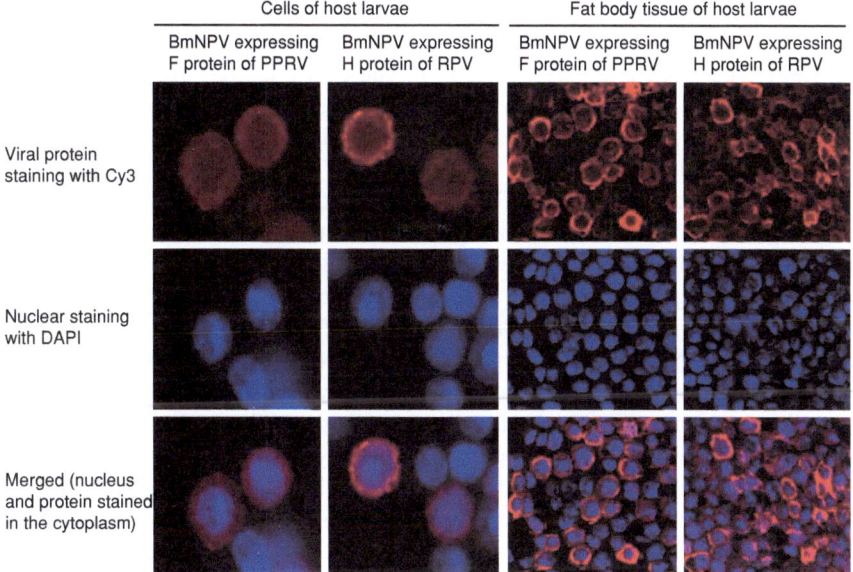

**Fig. 6.1** Immunolocalization of the BmNPV expressing F protein of PPRV or the H protein of RP in cells and fat body tissues of infected larvae, as monitored by indirect immunofluorescence. The localization of the recombinant proteins was detected by anti-F or anti-H antibodies followed by reaction with the Cy3-tagged secondary antibody (*orange*). The nuclei were stained with DAPI (*blue*). These images are reproduced from Rahman et al. (2003) with permission

**Table 6.2** Functional activities of antisera against the displayed proteins

| Property | – | Antibodies raised against | | |
|---|---|---|---|---|
| | | BmNPV expressing F protein of PPRV | BmNPV expressing H protein of RPV | Wild-type BmNPV |
| Virus neutralization titer[a] | PPRV (Nig 75/1) | 640 | 160 | 0 |
| | RPV (RBOK) | 640 | 320 | 0 |
| Hemagglutination inhibition titer[b] | – | 640 | 160 | 0 |

[a] Virus neutralization titer is defined as the reciprocal of the highest dilution of serum that neutralizes 50 % of virus infectivity

[b] Hemagglutination inhibition titer is defined as the reciprocal of the highest dilution of serum that inhibits hemagglutination activity of PPRV

Keeping large-scale production and potential of the F protein as a marker vaccine in mind, Singh and co-workers have tried to immunize goats either orally or by injecting extracts of larvae expressing F proteins. Unfortunately, neither the antigen nor the antibodies against PPRV were detected. The serum obtained from these goats also failed to be neutralizing for PPRV for at least 56 days post-immunization. It was latter speculated that since the F protein primarily elicits cell-meditated immune responses (Diallo et al. 2007), determination of this would have been a better criterion for immune status of the goats (Sen et al. 2010).

There have been immense efforts to rescue PPRV entirely from cloned DNA (infectious clone). However, there is no such clone available that can be used as a marker vaccine. Therefore, already established RPV infectious clones have been manipulated and exploited to be used as a PPRV marker vaccine. Initially, Das et al. (2000) created a chimeric RPV in which either the F or HN gene, or both, were replaced with the corresponding gene from the PPRV. Interestingly, it was noticed that the integrated activities of both surface glycoproteins (F and HN) are crucial for viral growth, since the chimeric RPV virus in which only one gene was swapped did not grow in the cell-culture system. The rescued virus expressing the PPRV F and HN glycoproteins grew more slowly in tissue culture than either parental virus and, formed abnormally large syncytia. Regardless of their growth in cell culture, goats infected with the chimera showed no adverse reaction, as assessed by clinical signs, temperature, leukocyte count, virus isolation and serology, and were protected from subsequent challenge with the wild-type PPRV. Based on these and other comprehensive results, it was concluded that these chimeric viruses can be used as a genetically marked vaccine, which will be useful for the control of PPRV, requiring epidemiological seromonitoring of PPRV prevalence and spread in the presence of vaccination (Das et al. 2000).

In order to improve the poor growth of the chimeric RPV in which only the F and HN were swapped with the corresponding genes of PPRV, Mahapatra et al. (2006) constructed a triple chimera virus that contained an additional M gene of PPRV. Considering that the poor growth may be due to nonhomologous interaction of the surface glycoproteins with the internal components of the virus, in particular with the M protein, the growth of the triple chimera was improved, and it grew to a titer as high as that of the unmodified PPRV, although comparatively lower than that of the parental RPV virus. As with the dual chimera (F and HN), this chimera virus did not cause any adverse reaction in immunized goats and the goats were protected from subsequent challenge with wild-type PPRV (Mahapatra et al. 2006). In the following year, the same group rescued a chimeric RPV that expressed the nucleocapsid protein derived from PPRV and suggested to use it as a marker vaccine (Parida et al. 2007).

The combined results of these studies indicated that devising a test based on the monoclonal antibodies response to the HN and N protein of RPV and PPRV could be applied for DIVA strategies (Libeau et al. 1994; Libeau et al. 1995). Additionally, using the same approach, it is possible to differentiate the vaccinated animals that subsequently become infected with either of the diseases (Mahapatra et al. 2006). However, the monoclonal antibodies used in available competitive ELISAs are cross-reacting with both RPV and PPRV. Therefore, this assay limits the application of marker vaccines in the field, and this fact emphasizes the need for a better companion test to be developed. Although the N protein has been used largely to design diagnostic assays, the C-terminus of the N protein appears to be highly variable among the morbilliviruses, and has been reported to protrude from the surface of the viral nucleocapsid (Heggeness et al. 1981). Therefore, it remains a suitable candidate for developing a test that can differentiate RPV from PPRV. A substantial contribution was recently made when the C-terminal variable region of the RPV N protein was expressed in *Escherichia coli*, and used subsequently to develop an indirect ELISA that could be used for serological identification of animals vaccinated with the chimeric vaccine (Parida et al. 2007).

Recently, an effort has been made to improve the thermotolerance of the vaccine while still maintaining the immunogenicity of the vaccine against PPRV. They have expressed a HN protein of PPRV in peanut plants (*Arachis hypogea*) in a naïve and biologically active format. It was interesting to observe that the expressed protein in the peanut plants retained its immunodominant epitopes in their natural conformation. The immunogenicity of the plant derived HN protein was analyzed in sheep upon oral immunization. Virus neutralizing antibody responses were elicited upon immunization of the sheep in the absence of any mucosal adjuvant. In addition, anti-PPRV-HN specific cell-mediated immune responses were also detected in mucosally immunized sheep (Khandelwal et al. 2011).

## 6.5 Multivalent Vaccines

PPRV, being significantly immunosuppressive, has been identified with other concurrent infections, notably bluetongue virus (BTV) (Mondal et al. 2009), sheep pox virus (SPV), goat pox virus (GPV) (Saravanan et al. 2007), and pestivirus (Kul et al. 2008). Because the geographical distribution of some of these diseases, such as GPV and PPRV, is similar, and developing countries demand economical infrastructures to support concerted vaccination programs, it is a requirement of the time to design multivalent (bi- or trivalent) vaccines. Additionally, field application of these multivalent vaccines to control common pathogens would, indeed, help to enhance poverty alleviation.

There have been great improvements in the development of human multivalent vaccines. There is a common use of tetravalent vaccines including measles, mumps, rubella, and varicella in children (Swartz et al. 1974). This highly advantageous form of vaccine has largely been ignored in veterinary medicine; however, only a few multivalent vaccines are available for pets and poultry. In canines, there is an available multivalent vaccine that is protective against canine distemper virus, canine adenovirus type 2, canine parvovirus type 2b, and canine parainfluenza viruses (Abdelmagid et al. 2004). Similarly, besides others, a freeze-dried vaccine that contains modified live virus strains of infectious bovine rhinotracheitis, bovine viral diarrhea, parainfluenza 3 and bovine respiratory syncytia viruses is currently being marketed by Pfizer under the trade name of BOVI-SHIELD® (Pfizer Animal Health).

Considering the enormous benefits of multivalent vaccines, currently vaccines are being developed that may both protect vaccinated animals against several viral pathogens and enable vaccinated and infected animals to be distinguished using

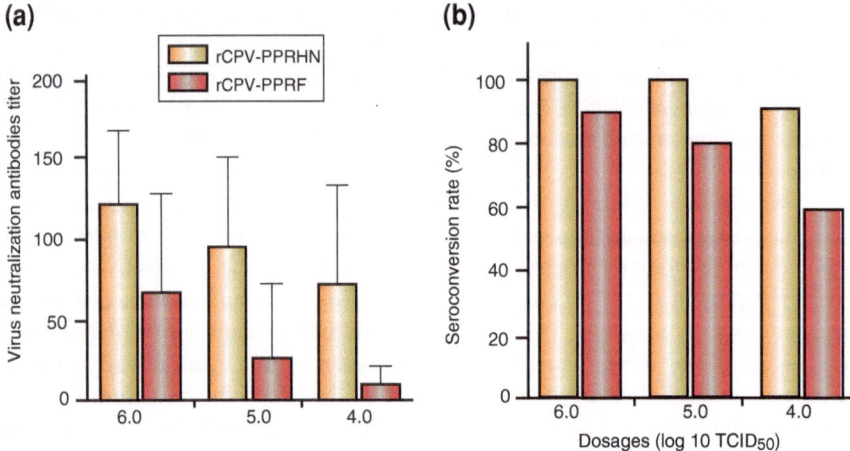

**Fig. 6.2** Virus neutralization **a** and seroconversion **b** responses in goats inoculated with rCPV-PPRH and rCPV-PPRF at different dosages. These figures are modified from Chen et al. (2010) with permission

DIVA tests for PPRV. As discussed earlier, PPRV is an enveloped RNA virus with two external glycoproteins, the F and HN proteins, which determine the protective immune status of animals against PPRV infection. This leaves the possibility that expression of these glycoproteins (F and HN) in various vector systems can be used as effective subunit vaccines. Following this approach, a dual recombinant vaccine was developed to protect small ruminants against PPRV and capripoxvirus infections at the same time. Sheep pox and goat pox are contagious diseases of small ruminants, especially of sheep and goats, respectively. These are primarily characterized by fever, lachrymation, and secondary bronchopneumonia with nasal discharges. Another member of the same virus group causes infection in cattle (lumpy skin disease). Effective attenuated vaccines are already available to control capripoxvirus infections (Kitching et al. 1987).

It is, therefore, plausible to use poxvirus as a vaccine vector to administer immunogenic genes from PPRV, which share the same geographical distribution. Initially, Diallo et al. (2002) reported a recombinant capripoxvirus that expresses the HN protein of PPRV. They showed that goats immunized with this recombinant capripoxvirus with a dose of at least $10$ $TCID_{50}$ are protective against virulent PPRV (Diallo et al. 2002). In the following year, the F gene of PPRV was incorporated into the backbone of a capripoxvirus from the same group. Immunization of animals with such a vaccine has shown protection to both economically important diseases (PPRV and GPV) with as low as 0.1 PFU of recombinant capripoxvirus vaccine (Berhe et al. 2003). Comparison of both these vaccines demonstrated that recombinant capripoxvirus expressing the HN protein requires a higher dose (about 100 times) than expressing the F protein to protect small ruminants against virulent challenge of PPRV. Although both recombinant viruses provided sound protection in challenge studies even with very low vaccine doses, their abilities to neutralize PPRV and subsequent inhibition of viral secretion were not determined.

Recently, in a comprehensive study, the potential roles of both surface glyco-proteins (F and HN) were investigated, again in a capripoxvirus-vectored vaccine (Chen et al. 2010). The authors have synthesized two recombinant capripoxviruses (rCPV) expressing either the F protein (rCPV-PPRVF) or the HN protein (rCPV-PPRVHN) of PPRV. Vaccination studies with different dosages of recombinant viruses showed that rCPV-PPRVHN were more potent inducer of PPRV virus neutralization antibodies (VNA) than rCPV-PPRVF (Fig. 6.2a). One dose of rCPV-PPRVHN was enough to seroconvert 80 % of the immunized sheep (Fig. 2b). A second dose induced significantly higher PPRV VNA titers. There was no significant difference in PPRV VNA responses between goats and sheep. Moreover, vaccination with rCPV-PPRVHN also protected goats from virulent capripoxvirus challenge (Chen et al. 2010). They proposed that this vaccine could be a practical and useful candidate DIVA vaccine, accompanied by ELISA against N protein, in countries where PPR is newly emerging or where stamp-out plans are yet to be implemented.

Correspondingly, a Vero cell has been developed in which vaccines against sheep pox (Romanian Fanar strain) and PPR (Sungri/96 strain) are combined. This dual-vaccine was found to protect small ruminants against both PPR and sheep pox, simultaneously, as was evident from seroconversion as well as protection on homologous challenge in sheep, implying that the vaccine viruses did not interfere with the immunogenicity of each other (Chaudhary et al. 2009). Single immuni-zation with live PPR vaccine has been able to maintain protective levels of serum antibody for up to 4 years, while sheep pox vaccine was found to confer protective immunity for at least 2 years. Single vaccination covering both diseases can facilitate greater convenience, significantly bring down the cost of vaccination, and reduce stress to the animals, and also to vaccination teams.

Efforts have been made to estimate the level and longevity of protection of multivalent vaccines in field conditions. One such effort was made in Cameroon when 20 goats were immunized with a mixed vaccine carrying attenuated PPRV 75/l strain for PPR and the RM 65 strain for capripoxvirus, and then challenged with wild virulent strains of PPR and goat pox (Martrenchar et al. 1997). Unfor-tunately, the RM 65 strain did not act as an effective vaccine against the virulent field strain of goat pox used in the challenge experiment. The authors speculated that the failure of these vaccines could be due to a lack of cross protection between the RM 65 strain and the challenged wild goat pox strain, or due to the RM 65 vaccine adaptation to the Vero cell culture (Martrenchar et al. 1997). Based on these observations, and partial protection between sheep and goat pox viruses, a homologous vaccine was recommended for full protection (Bhanuprakash et al. 2006). Contrary to this report, a combined vaccine for PPRV goat pox virus induced a protective immune response and sustained the homologous challenge in goats (Hosamani et al. 2006). These discrepancies need to be studied at the molecular level, despite the fact that both viruses (PPRV Sungri 1996 and GTPV Uttarkashi 1978) used in the latter study were different from the former study.

Nevertheless, these provided evidence that PPRV and goat pox viruses do not interfere in each other's immunogenicity, and may be suitable bivalent vaccines in the region where both these diseases are prevalent. However, the duration of immunity conferred by the combined vaccine still remains to be determined.

## 6.6  DIVA Vaccines

The available live attenuated vaccine against PPRV is highly successful and is currently being used in most of the PPRV endemic countries. However, this vaccine cannot be used for the longer term, as it would interfere with disease surveillance based on serological testing, and may result in the loss of a country's disease-free status. The major reason for this is that the antibody responses in animals immunized with this live attenuated vaccine cannot be distinguished from those following a natural infection. Ultimately, the seromonitoring of the disease will be unmanageable, especially in the areas where the disease is endemic and either the vaccination has been implemented or is being implemented. Having in mind that vaccination is essential to control the disease and that serosurveillance is crucial to estimate the disease prevalence, it is inevitable to design such a vaccine that is protective against the disease and has the ability to be differentiated from naturally infected animals. An acronym is used for such a vaccine: differentiation between infected and vaccinated animals (DIVA).

There have been immense efforts to rescue PPRV entirely from cloned DNA (infectious clone) in the lab of Prof. Baron (formerly Prof. Barret's lab). Currently, there have been some claims that it was possible to rescue PPRV strain Nig/75/1 (Satya Parida, Personal communication). However, taking advantage of available RPV infectious clones, several chimeric constructs have been prepared in which different genes of PPRV and RPV were interchanged, and their immunogenicity and protection were determined, with the ultimate goal of improved marker and DIVA vaccine, as discussed earlier. Recently, Buczkowski et al. (2012) have described a novel mechanism of marking morbillivirus vaccines, using RPV as a proof of concept, and they discuss the applicability of this method to the development of marked vaccines for PPRV (Buczkowski et al. 2012).

Despite the fact of significant improvement in recent years, no concrete DIVA system is currently available. Therefore, there is a need to identify and selectively delete genes from the PPRV genome and to build an infectious clone. It should be expected that this clone will allow the development of "marker vaccines" that, combined with suitable diagnostic assays, allow differentiation of infected from vaccinated animals by different antibody responses induced by the vaccine (no antibodies generated to deleted genes) from those induced during infection with the wild-type virus. Such DIVA vaccines and their companion diagnostic tests are now available or in development for several diseases, including infectious bovine rhinotracheitis, pseudorabies, classical swine fever, and foot and mouth disease, among others (Meeusen et al. 2007).

## 6.7   Control of PPRV Using RNA Interference

Recently, a new method has been developed to control morbilliviruses included PPRV based on a novel technique derived from molecular genetics (Servan de Almeida et al. 2007). This approach targets a natural biological mechanism of eukaryotic cells known as RNA interference (RNAi). This RNAi allows multicellular organisms to control the level of expression of some of their genes. The process involves short RNA fragments capable of preventing the reading and translation into proteins of the genetic code carried by DNA: the fragments are known as interfering RNA. They prevent the RNA playing its fundamental role as a messenger of the information contained in the genes with a view to protein production. In effect, so-called interfering RNA links specifically to the target messenger RNA, resulting in the latter's deterioration and consequently inhibiting expression of the corresponding protein. It has been shown that three synthetic RNAi targeting N gene of PPRV inhibits more than 80 % of virus replication in vitro. They are targeted at the messenger RNA of the N gene of the viruses that cause the diseases, blocking the virus multiplication process. The next step is to evaluate the antiviral actions of these RNAi in vivo. It is expected that novel strategy would open the way for the development of therapeutic vaccines against PPRV and should make it possible to provide farmers with a safe and effective vaccine (Servan de Almeida et al. 2007).

## 6.8   Conclusions

There has been a substantial improvement in the detection of nucleic acids of PPRV in recent years. Demonstration of several real-time PCR assays has provided powerful and novel means of not only detection but also quantification of nucleic acid of PPRV in several kinds of clinical samples. However, these diagnostic tools are not readily available in all diagnostic laboratories, especially in developing nations. There is a need to establish reliable, sensitive, and affordable diagnostic tools, which will be promptly accessible at low cost, and independent of laboratory type. Therefore, it is observed that there is a strong tendency to increase the number of diagnostic tools that are based on diverse principles. While doing so, it is extremely important to design these assays in a way that will require less time, should be readily available, affordable for developing nations, not requiring high-tech facilities in laboratories, and must not be complex while being performed in field conditions. Although most of the lineages are continent specific, reports on mixed lineages are emerging such as in Sudan and Uganda. None of the available assays is devised to differentiate all of the lineages, which warrants further investigations.

Although veterinary vaccines contribute only with around 23 % of animal health products, the trend is increasing tremendously. There is always an element of encouragement to use veterinary vaccines, because of their favorable impact on public health through reductions in the use of veterinary pharmaceuticals and hormones and their residues in the human food chain. In addition, vaccines contribute to the well-being of livestock and companion animals; and the growing animal welfare lobby favors their use. In this respect, there is a huge requirement for the PPRV vaccines to adapt next-generation vaccine strategies with the ultimate aim to develop cheap, efficient, and multivalent vaccines. The research is in progress in all these aspects of PPRV, and it is expected that an efficient system for diagnosis, immunization, and DIVA will be available soon.

# References

Abdelmagid OY, Larson L, Payne L, Tubbs A, Wasmoen T, Schultz R (2004) Evaluation of the efficacy and duration of immunity of a canine combination vaccine against virulent parvovirus, infectious canine hepatitis virus, and distemper virus experimental challenges. Vet Therapeutics Res Appl Vet Med 5(3):173–186

Adebayo AA, Sim-Brandenburg JW, Emmel H, Olaleye DO, Niedrig M (1998) Stability of 17D yellow fever virus vaccine using different stabilizers. Biologicals 26(4):309–316

Adu FD, Nawathe DR (1981) Safety of tissue culture rinderpest vaccine in pregnant goats. Trop Anim Health Prod 13(3):166

Anderson J, McKay JA (1994) The detection of antibodies against peste des petits ruminants virus in cattle, sheep and goats and the possible implications to rinderpest control programmes. Epidemiol Infect 112(1):225–231

Asim M, Rashid A, Chaudhary AH (2008) Effect of various stabilizers on titre of lyophilized live-attenuated Peste des Petits Ruminants (PPR) Vaccine. Pakistan Vet J 28(4):203–204

Balamurugan V, Sen A, Saravanan P, Singh RP, Singh RK, Rasool TJ, Bandyopadhyay SK (2006) One-step multiplex RT-PCR assay for the detection of peste des petits ruminants virus in clinical samples. Vet Res Commun 30(6):655–666

Bao J, Li L, Wang Z, Barrett T, Suo L, Zhao W, Liu Y, Liu C, Li J (2008) Development of one-step real-time RT-PCR assay for detection and quantitation of peste des petits ruminants virus. J Virol Methods 148(1–2):232–236

Berhe G, Minet C, Le Goff C, Barrett T, Ngangnou A, Grillet C, Libeau G, Fleming M, Black DN, Diallo A (2003) Development of a dual recombinant vaccine to protect small ruminants against peste-des-petits-ruminants virus and capripoxvirus infections. J Virol 77(2):1571–1577

Bhanuprakash V, Indrani BK, Hosamani M, Singh RK (2006) The current status of sheep pox disease. Comp Immunol Microbiol Infect Dis 29(1):27–60

Bodjo SC, Kwiatek O, Diallo A, Albina E, Libeau G (2007) Mapping and structural analysis of B-cell epitopes on the morbillivirus nucleoprotein amino terminus. J Gen Virol 88(Pt 4):1231–1242

Bonniwell MA (1980) The use of Tissue Culture Rinderpest Vaccine to protect sheep and goats against peste des petits ruminants in the Ashanti region of Ghana. Bull Off Int Epiz 92:1233–1238

Bourdin P, Rioche M, Laurent A (1970) Emploi d'un vaccin antibovipestique produit sur cultures cellulaires dans la prophylaxie de la peste des petits ruminants au Dahomey. Rev Elev Med Vet Pays Trop 23:295–300

Buczkowski H, Parida S, Bailey D, Barrett T, Banyard AC (2012) A novel approach to generating morbillivirus vaccines: negatively marking the rinderpest vaccine. Vaccine 30(11):1927–1935

Chaudhary SS, Pandey KD, Singh RP, Verma PC, Gupta PK (2009) A vero cell derived combined vaccine against sheep pox and Peste des Petits ruminants for sheep. Vaccine 27(19):2548–2553

Chen W, Hu S, Qu L, Hu Q, Zhang Q, Zhi H, Huang K, Bu Z (2010) A goat poxvirus-vectored peste-des-petits-ruminants vaccine induces long-lasting neutralization antibody to high levels in goats and sheep. Vaccine 28(30):4742–4750

Choi KS, Nah JJ, Ko YJ, Kang SY, Yoon KJ, Jo NI (2005) Antigenic and immunogenic investigation of B-cell epitopes in the nucleocapsid protein of peste des petits ruminants virus. Clin Diagn Lab Immunol 12(1):114–121

Cosby S, Kai C, Yamanouchi K (2005) Immunology of rinderpest- an immunosuppression but a lifelong vaccine protection. Rinderpest and peste des petits ruminants. Virus plagues of large and small ruminants. Academic Press, Elsevier, Amsterdam

Couacy-Hymann E, Roger F, Hurard C, Guillou JP, Libeau G, Diallo A (2002) Rapid and sensitive detection of peste des petits ruminants virus by a polymerase chain reaction assay. J Virol Methods 100:17–25

Das SC, Baron MD, Barrett T (2000) Recovery and characterization of a chimeric rinderpest virus with the glycoproteins of peste-des-petits-ruminants virus: homologous F and H proteins are required for virus viability. J Virol 74(19):9039–9047

Devireddy LR, Raghavan R, Ramachandran S, Shaila MS (1999) The fusion protein of peste des petits ruminants virus is a hemolysin. Arch Virol 144(6):1241–1247

Devireddy LR, Raghavan R, Ramachandran S, Subbarao SM (1998) Protection of rabbits against lapinized rinderpest virus with purified envelope glycoproteins of peste-des-petits-ruminants and rinderpest viruses. Acta Virol 42(5):299–306

Dhinakar Raj G, Nachimuthu K, Mahalinga Nainar A (2000) A simplified objective method for quantification of peste des petits ruminants virus or neutralizing antibody. J Virol Methods 89(1–2):89–95

Diallo A (1990) Morbillivirus group: genome organisation and proteins. Vet Microbiol 23(1–4):155–163

Diallo A (2004) Peste des Petits Ruminants. In: Manual of diagnostic tests and vaccines for terrestrial animals. OIE

Diallo A, Barrett T, Barbron M, Meyer G, Lefevre PC (1994) Cloning of the nucleocapsid protein gene of peste-des-petits-ruminants virus: relationship to other morbilliviruses. J Gen virol 75(Pt 1):233–237

Diallo A, Libeau G, Couacy-Hymann E, Barbron M (1995) Recent developments in the diagnosis of rinderpest and peste des petits ruminants. Vet Microbiol 44(2–4):307–317

Diallo A, Minet C, Berhe G, Le Goff C, Black DN, Fleming M, Barrett T, Grillet C, Libeau G (2002) Goat immune response to capripox vaccine expressing the hemagglutinin protein of peste des petits ruminants. Ann N Y Acad Sci 969:88–91

Diallo A, Minet C, Le Goff C, Berhe G, Albina E, Libeau G, Barrett T (2007) The threat of peste des petits ruminants: progress in vaccine development for disease control. Vaccine 25(30):5591–5597

Diallo A, Taylor WP, Lefèvre PC, Provost A (1989) Atténuation d'une souche de virus de la peste des petits ruminants : candidat pour un vaccin homologue vivant. Rev Elev Med Vet Pays Trop 42:311–317

Eligulashvili R, Bumbarov V, Yadin H (2002) Comparison of the avidin-biotin and polymer detection systems for rapid recognition of peste des petits ruminants (PPR) virus in situ. Bulg J Vet Med 5(4):229–232

Forsyth MA, Barrett T (1995) Evaluation of polymerase chain reaction for the detection and characterisation of rinderpest and peste des petits ruminants viruses for epidemiological studies. Virus Res 39(2–3):151–163

Ghosh S, Parvez MK, Banerjee K, Sarin SK, Hasnain SE (2002) Baculovirus as mammalian cell expression vector for gene therapy: an emerging strategy. Mol Ther J Am Soc Gene Ther 6(1):5–11

Gibbs EP, Taylor WP, Lawman MJ, Bryant J (1979) Classification of peste des petits ruminants virus as the fourth member of the genus Morbillivirus. Intervirology 11(5):268–274

Gilbert Y, Monnier J (1962) Adaptation du virus de la peste des petits ruminants aux cultures cellulaires. Rev Elev Med Vet Pays Trop 15:321

Haffar A, Libeau G, Moussa A, Cecile M, Diallo A (1999) The matrix protein gene sequence analysis reveals close relationship between peste des petits ruminants virus (PPRV) and dolphin morbillivirus. Virus Res 64(1):69–75

Hamdy FM, Dardiri AH, Nduaka O, Breese SRJ, Ihemelandu EC (1976) Etiology of the stomatitis pneumcenteritis complex in Nigerian dwarf goats. Can J Comp Med 40:276–284

Heggeness MH, Scheid A, Choppin PW (1981) The relationship of conformational changes in the Sendai virus nucleocapsid to proteolytic cleavage of the NP polypeptide. Virology 114(2):555–562

Hosamani M, Singh SK, Mondal B, Sen A, Bhanuprakash V, Bandyopadhyay SK, Yadav MP, Singh RK (2006) A bivalent vaccine against goat pox and Peste des Petits ruminants induces protective immune response in goats. Vaccine 24(35–36):6058–6064

Ismail TM, Yamanaka MK, Saliki JT, el-Kholy A, Mebus C, Yilma T (1995) Cloning and expression of the nucleoprotein of peste des petits ruminants virus in baculovirus for use in serological diagnosis. Virology 208(2):776–778

Jones L, Giavedoni L, Saliki JT, Brown C, Mebus C, Yilma T (1993) Protection of goats against peste des petits ruminants with a vaccinia virus double recombinant expressing the F and H genes of rinderpest virus. Vaccine 11(9):961–964

Keerti M, Sarma BJ, Reddy YN (2009) Development and application of latex agglutination test for detection of PPR virus. Indian Vet J 86:234–237

Khandelwal A, Renukaradhya GJ, Rajasekhar M, Sita GL, Shaila MS (2011) Immune responses to hemagglutinin-neuraminidase protein of peste des petits ruminants virus expressed in transgenic peanut plants in sheep. Vet Immunol Immunopathol 140(3–4):291–296

Kitching RP, Hammond JM, Taylor WP (1987) A single vaccine for the control of capripox infection in sheep and goats. Res Vet Sci 42(1):53–60

Kul O, Kabakci N, Ozkul A, Kalender H, Atmaca HT (2008) Concurrent peste des petits ruminants virus and pestivirus infection in stillborn twin lambs. Vet Pathol 45(2):191–196

Kwiatek O, Keita D, Gil P, Fernandez-Pinero J, Jimenez Clavero MA, Albina E, Libeau G (2010) Quantitative one-step real-time RT-PCR for the fast detection of the four genotypes of PPRV. J Virol Methods 165(2):168–177

Lamb RA (1993) Paramyxovirus fusion: a hypothesis for changes. Virology 197(1):1–11

Laurent A (1968) Aspects biologiques de la multiplication du virus de la peste des petits ruminants ou PPR sur cultures cellulaires. Rev Elev Med Vet Pays Trop 21(3):297–308

Lefevre PC, Diallo A, Schenkel F, Hussein S, Staak G (1991) Serological evidence of peste des petits ruminants in Jordan. Vet Rec 128(5):110

Li L, Bao J, Wu X, Wang Z, Wang J, Gong M, Liu C, Li J (2010) Rapid detection of peste des petits ruminants virus by a reverse transcription loop-mediated isothermal amplification assay. J Virol Methods 170(1–2):37–41

Libeau G, Diallo A, Colas F, Guerre L (1994) Rapid differential diagnosis of rinderpest and peste des petits ruminants using an immunocapture ELISA. Vet Rec 134(12):300–304

Libeau G, Prehaud C, Lancelot R, Colas F, Guerre L, Bishop DH, Diallo A (1995) Development of a competitive ELISA for detecting antibodies to the peste des petits ruminants virus using a recombinant nucleoprotein. Res Vet Sci 58(1):50–55

Mahapatra M, Parida S, Baron MD, Barrett T (2006) Matrix protein and glycoproteins F and H of Peste-des-petits-ruminants virus function better as a homologous complex. J Gen Virol 87(Pt 7):2021–2029

Manoharana S, Jayakumarb R, Govindarajanc R, Koteeswarana A (2005) Haemagglutination as a confirmatory test for Peste des petits ruminants diagnosis. Small Rumin Res 59(1):75–78

Mariner JC, House JA, Sollod AE, Stem C, van den Ende M, Mebus CA (1990) Comparison of the effect of various chemical stabilizers and lyophilization cycles on the thermostability of a Vero cell-adapted rinderpest vaccine. Vet Microbiol 21(3):195–209

Martrenchar A, Zoyem N, Diallo A (1997) Experimental study of a mixed vaccine against peste des petits ruminants and capripox infection in goats in Northern Cameroon. Small Rumin Res 26:39–44

McCullough KC, Sheshberadaran H, Norrby E, OBI TU, Crowther JR (1986) Monoclonal antibodies against morbilliviruses. Rev sci Tech Off Int Epiz 5(2):411–427

Meeusen EN, Walker J, Peters A, Pastoret PP, Jungersen G (2007) Current status of veterinary vaccines. Clin Microbiol Rev 20(3):489–510

Mondal B, Sen A, Chand K, Biswas SK, De A, Rajak KK, Chakravarti S (2009) Evidence of mixed infection of peste des petits ruminants virus and bluetongue virus in a flock of goats as confirmed by detection of antigen, antibody and nucleic acid of both the viruses. Trop Anim Health Prod 41(8):1661–1667

Munir M (2011) Diagnosis of Peste des Petits Ruminants under limited resource setting: A cost effective strategy for developing countries where PPRV is endemic. VDM Verlag Dr, Müller, Germany

Munir M, Abubakar M, Khan MT, Abro SH (2009a) Comparative efficacy of single radial haemolysis test and countercurrent immunoelectro-osmo-phoresis with monoclonal antibodies-based com-petitive elisa for the serology of peste des petits ruminants in sheep and goats. Bulg J Vet Med 12(4):246–253

Munir M, Abubakar M, Zohari S, Berg M (2012a) Serodiagnosis of Peste des Petits Ruminants Virus. In: Al-Moslih M (ed) Serological diagnosis of certain human, animal and plant diseases, vol 1. InTech, pp 37–58

Munir M, Siddique M, Ali Q (2009b) Comparative efficacy of standard AGID and precipitinogen inhibition test with monoclonal antibodies based competitive ELISA for the serology of Peste des Petits Ruminants in sheep and goats. Trop Anim Health Prod 41(3):413–420

Munir M, Zohari S, Saeed A, Khan QM, Abubakar M, LeBlanc N, Berg M (2012b) Detection and phylogenetic analysis of peste des petits ruminants virus isolated from outbreaks in punjab, pakistan. Transboundary Emerg Dis 59(1):85–93

Munir M, Zohari S, Suluku R, Leblanc N, Kanu S, Sankoh FA, Berg M, Barrie ML, Stahl K (2012c) Genetic characterization of peste des petits ruminants virus, sierra leone. Emerg Infect Dis 18(1):193–195

Muthuchelvan D, Sanyal A, Sreenivasa BP, Saravanan P, Dhar P, Singh RP, Singh RK, Bandyopadhyay SK (2006) Analysis of the matrix protein gene sequence of the Asian lineage of peste-des-petits ruminants vaccine virus. Vet Microbiol 113(1–2):83–87

Nagamine K, Hase T, Notomi T (2002) Accelerated reaction by loop-mediated isothermal amplification using loop primers. Mol Cell Probes 16(3):223–229

Norrby E, Sheshberadaran H, McCullough KC, Carpenter WC, Orvell C (1985) Is rinderpest virus the archevirus of the Morbillivirus genus? Intervirology 23(4):228–232

Obi TU (1984) The detection of PPR virus antigen by agar gel precipitation test and counter-immunoelectrophoresis. J Hyg 93:579–586

Parida S, Mahapatra M, Kumar S, Das SC, Baron MD, Anderson J, Barrett T (2007) Rescue of a chimeric rinderpest virus with the nucleocapsid protein derived from peste-des-petits-ruminants virus: use as a marker vaccine. J Gen Virol 88(Pt 7):2019–2027

Plowright W, Ferris RD (1959) Studies with rinderpest virus in tissue culture. II. Pathogenicity for cattle of culture-passaged virus. J Comp Pathol 69(2):173–184

Rahman MM, Shaila MS, Gopinathan KP (2003) Baculovirus display of fusion protein of Peste des petits ruminants virus and hemagglutination protein of Rinderpest virus and immunogenicity of the displayed proteins in mouse model. Virology 317(1):36–49

Rajak KK, Sreenivasa BP, Hosamani M, Singh RP, Singh SK, Singh RK, Bandyopadhyay SK (2005) Experimental studies on immunosuppressive effects of peste des petits ruminants (PPR) virus in goats. Comp Immunol Microbiol Infect Dis 28(4):287–296

Ramachandran S, Shaila MS (1995) Shyam G Hemagglutination and hemadsorption by peste des petits ruminants virus (PPRV). Immunobiology of viral infections. In: Schwayzer M, Ackermann M, Bertone G et al (eds) Third Congress Eur. Soc. Vet. Virol, Switzerland, pp 513–515

Romero CH, Barrett T, Kitching RP, Bostock C, Black DN (1995) Protection of goats against peste des petits ruminants with recombinant capripoxviruses expressing the fusion and haemagglutinin protein genes of rinderpest virus. Vaccine 13(1):36–40

Rossiter PB (2004) Peste des petits ruminants. In: Coetzer J (ed) Infectious disease of livestock. 2nd edn. Oxford University Press, South Africa, pp 660–672

Saliki JT, House JA, Mebus CA, Dubovi EJ (1994) Comparison of monoclonal antibody-based sandwich enzyme-linked immunosorbent assay and virus isolation for detection of peste des petits ruminants virus in goat tissues and secretions. J Clin Microbiol 32(5):1349–1353

Saliki JT, Libeau G, House JA, Mebus CA, Dubovi EJ (1993) Monoclonal antibody-based blocking enzyme-linked immunosorbent assay for specific detection and titration of peste-des-petits-ruminants virus antibody in caprine and ovine sera. J Clin Microbiol 31(5):1075–1082

Saravanan P, Balamurugan V, Sen A, Sarkar J, Sahay B, Rajak KK, Hosamani M, Yadav MP, Singh RK (2007) Mixed infection of peste des petits ruminants and orf on a goat farm in Shahjahanpur. India Vet Rec 160(12):410–412

Saravanan P, Sen A, Balamurugan V, Bandyopadhyay SK, Singh RK (2008) Rapid quality control of a live attenuated Peste des petits ruminants (PPR) vaccine by monoclonal antibody based sandwich ELISA. Biolog J Int Assoc Biolog Stand 36(1):1–6

Saravanan P, Sen A, Balamurugan V, Rajak KK, Bhanuprakash V, Palaniswami KS, Nachimuthu K, Thangavelu A, Dhinakarraj G, Hegde R, Singh RK (2010) Comparative efficacy of peste des petits ruminants (PPR) vaccines. Biolog J Int Assoc Biolog Stand 38(4):479–485

Sarkar J, Sreenivasa BP, Singh RP, Dhar P, Bandyopadhyay SK (2003) Comparative efficacy of various chemical stabilizers on the thermostability of a live-attenuated peste des petits ruminants (PPR) vaccine. Vaccine 21(32):4728–4735

Sen A, Saravanan P, Balamurugan V, Rajak KK, Sudhakar SB, Bhanuprakash V, Parida S, Singh RK (2010) Vaccines against peste des petits ruminants virus. Expert Rev Vaccines 9(7):785–796

Servan de Almeida R, Keita D, Libeau G, Albina E (2007) Control of ruminant morbillivirus replication by small interfering RNA. J Gen Virol 88(Pt 8):2307–2311

Seth S, Shaila MS (2001) The hemagglutinin-neuraminidase protein of peste des petits ruminants virus is biologically active when transiently expressed in mammalian cells. Virus Res 75(2):169–177

Sharma B, Norrby E, Blixenkrone-Moller M, Kovamees J (1992) The nucleotide and deduced amino acid sequence of the M gene of phocid distemper virus (PDV). The most conserved protein of morbilliviruses shows a uniquely close relationship between PDV and canine distemper virus. Virus Res 23(1–2):13–25

Silva AC, Carrondo MJ, Alves PM (2011) Strategies for improved stability of Peste des Petits Ruminants Vaccine. Vaccine 29(31):4983–4991

Silva AC, Delgado I, Sousa MF, Carrondo MJ, Alves PM (2008) Scalable culture systems using different cell lines for the production of Peste des Petits ruminants vaccine. Vaccine 26(26):3305–3311

Singh RP, Sreenivasa BP, Dhar P, Bandyopadhyay SK (2004a) A sandwich-ELISA for the diagnosis of Peste des petits ruminants (PPR) infection in small ruminants using anti-nucleocapsid protein monoclonal antibody. Arch Virol 149(11):2155–2170

Singh RP, Sreenivasa BP, Dhar P, Shah LC, Bandyopadhyay SK (2004b) Development of a monoclonal antibody based competitive-ELISA for detection and titration of antibodies to peste des petits ruminants (PPR) virus. Vet Microbiol 98(1):3–15

Sinnathamby G, Seth S, Nayak R, Shaila MS (2004) Cytotoxic T cell epitope in cattle from the attachment glycoproteins of rinderpest and peste des petits ruminants viruses. Viral Immunol 17(3):401–410

Sreenivasa BP, Dhar P, Singh RP, Bandyopadhyay SK (2002) Development of peste des petits ruminants challange virus from a field isolate. In: Annual converence and national seminar on management of viral diseases with emphasis on global trade and WTO regime of indian virological society, Hebbal, Bangalore, India, 18–20

Steinhauer DA, de la Torre JC, Holland JJ (1989) High nucleotide substitution error frequencies in clonal pools of vesicular stomatitis virus. J Virol 63(5):2063–2071

Sumption KJ, Aradom G, Libeau G, Wilsmore AJ (1998) Detection of peste des petits ruminants virus antigen in conjunctival smears of goats by indirect immunofluorescence. Vet Rec 142(16):421–424

Swartz TA, Klingberg W, Klingberg MA (1974) Combined trivalent and bivalent measles, mumps and rubella virus vaccination. A control Trial Infection 2(3):115–117

Taylor WP, Abegunde A (1979) The isolation of peste des petits ruminants virus from Nigerian sheep and goats. Res Vet Sci 26(1):94–96

Wei L, Gang L, Fan XJ, Zhang K, Jia FQ, Shi LJ, Unger H (2009) Establishment of a rapid method for detection of peste des petits ruminants virus by a reverse transcription loop-mediated isothermal amplification. Chin J Prev Vet Med 31(5):374–378

Worrall EE, Litamoi JK, Seck BM, Ayelet G (2000) Xerovac: an ultra rapid method for the dehydration and preservation of live attenuated Rinderpest and Peste des Petits ruminants vaccines. Vaccine 19(7–8):834–839

Wu R, Georgescu MM, Delpeyroux F, Guillot S, Balanant J, Simpson K, Crainic R (1995) Thermostabilization of live virus vaccines by heavy water (D2O). Vaccine 13(12):1058–1063

Yadav V, Balamurugan V, Bhanuprakash V, Sen A, Bhanot V, Venkatesan G, Riyesh T, Singh RK (2009) Expression of Peste des petits ruminants virus nucleocapsid protein in prokaryotic system and its potential use as a diagnostic antigen or immunogen. J Virol Methods 162(1–2):56–63

# Chapter 7
# Poverty Alleviation and Global Eradication of Peste des Petits Ruminants

**Abstract** Given its impact on animal health and its economic relevance, Peste des Petits ruminants (PPR) is an Office International des Epizooties (OIE) list A disease. The economic losses associated with PPR occur not only directly through reduced animal production and high death rate but also indirectly through trade losses due to restrictions on animal movements. The costs of implementing control measures along with diagnostic tests further influence the profitability of the small ruminant business. Recent outbreaks of PPRV in Morocco and Turkey highlight the importance of the disease and reasoned to facilitate the efforts for eradication of PPR, globally. Recently, rinderpest (RP) has been eradicated from the globe and efforts are in progress to eradicate other viral diseases, of which PPR is the most suitable candidate. The availability of efficient diagnostic tests accompanied by vaccines providing strong immunity that last for years are facilitating elements in this cause. However, a unified framework is currently lacking that can help bridging and synthesizing the lessons from the RP eradication programme, and effectively devise future campaigns for the successful control and elimination of PPR from the globe. Expecting the impact of great economic reward, a high number of PPR-endemic countries should join the force, and implement regional roadmaps for the progressive and successful control and elimination of PPRV. This chapter focused on all of these possibilities in light of the global concern and animal health organization's objectives. Moreover, efforts are made to emphasize the elements that favor the eradication campaign, while the research and planning gaps that require immediate attention are critically reviewed.

**Keywords** Control · Eradication · Planning · Economics · Strategies for animal disease elimination

M. Munir et al., *Molecular Biology and Pathogenesis of Peste des Petits Ruminants Virus*, SpringerBriefs in Animal Sciences, DOI: 10.1007/978-3-642-31451-3_7, © The Author(s) 2013

## 7.1 Introduction

The recent announcement by the global scientific community of the successful
worldwide eradication of rinderpest virus (RPV) is providing a renewed drive to
focus on ways of controlling its cousin PPRV, a similar disease that is increasingly
threatening Asian and African small ruminant and camel populations. In recent
years, frequent and severe shocks to production systems, including floods and
droughts, have had serious impacts on semi-nomadic areas. These negative effects
have been further aggravated by outbreaks of PPRV in many endemic countries.
The great economic impact, either directly or indirectly, has further necessitated
the eradication of PPR. Following the model of RP eradication, it seems logical
and feasible to eradicate PPR likewise; however, the scientific community is not
yet fully equipped for this challenge. Still, there are several natural and acquired
elements that favor the possibility of PPR eradication in the near future. How
much are we prepared and what shortcomings exist in PPR research are discussed
in this chapter.

## 7.2 Economic Impact of PPR Disease

Besides health complications and costs, animal diseases impose a wide range of
both direct and indirect loses. In estimating the economic impact of PPR in small
ruminants, the direct costs of disease incidence and control should in particular be
considered on specific stakeholders, but the indirect costs, including losses in the
food supply chain and the losses associated with households and enterprises,
should also be evaluated. Regarding PPRV, the factors that mediate these losses
are not well understood, and not thoroughly evaluated and estimated. In our
experience, the indirect costs associated with PPRV much outweigh the direct
ones. Although there are many parameters available to facilitate the evaluation of
the economic impact of a disease, there are several drawbacks to applying these,
such as the fact that they are subject specific and handle only one factor at a time,
and therefore lack the ability to estimate the cumulative impact on the economy.
Nevertheless, these economy-wide considerations are crucial in planning control
and eradication strategies for any emerging disease. It is then advised to conduct a
well-planned cost-benefit analysis of PPR verses policy responses that include both
the direct and indirect impacts associated with PPRV.

Since RP is eradicated, perhaps the most worrying thing for farmers in many
parts of Asia and Africa is the reality of another highly contagious viral disease
(PPR) that affects goats and sheep, gives similar signs to rinderpest, and is
becoming more widespread. Due to its highly contagious and acute nature, PPR
causes mortality up to 100 % in small ruminants, which substantially contributes
to the economics of small rural farms, especially where sheep and goats are reared
as the sole source of income. The disease is mostly present in developing

countries, which often rely heavily on subsistence farming of small ruminants for trade and food supply. As a consequence, the economic impact of PPR, especially in a naïve population, can be devastating. This situation becomes even more complicated when PPRV is confused with other diseases. Although experienced veterinarians may diagnose the disease based on clinical signs, similarity of the clinical picture to that of other respiratory diseases can present problems for differential diagnosis (see Sect. 3.3). Due to this error in PPRV diagnosis, the economic consequences of the disease in small ruminants remained underestimated (Taylor 1984). Despite the fact that PPR has been restricted to Africa, Asia, and the Middle East, it has expanded in past 10 years (Wang et al. 2009; Kwiatek et al. 2007; FAO September 9, 2008; Banyard et al. 2010) (see Chap. 5). Currently, about 62.5 % (1 billion) of the global small ruminant population is at risk of infection with PPRV [FAO sheet ref 33, described in (Arzt et al. 2010)].

Although several studies have considered that PPR is the major constraint for small ruminant production, the economic impact of PPR has not been fully evaluated (Rossiter and Taylor 1994; Ezeokoli et al. 1986; Nanda et al. 1996). In 1993, (Stem 1993) reported a macro-economic study in which government of Niger estimated the impact of PPR vaccination in one million head. The assumption made for the 5-year demographic model indicated that PPR vaccination is highly beneficial, with an anticipated return of 24 million dollars over 2 million dollars investment.

Another study, conducted by Awa et al. (2000) in 1996–1997 based on 18,400 head, indicated that the profit can be increased two- to threefold fold in goats and three- to fourfold in sheep by PPR vaccination and strategic antiparasitic treatment, considering that the PPR and helminthic infestation are the main constraints on production (Awa et al. 2000). The importance of PPR was realized in Asia and Africa when an international study was conducted to prioritize the importance of research. In this report, Perry et al. (2002) placed PPR among the top 10 diseases in sheep and goats, as these have high impact on the poor rural small ruminant farmers (Perry et al. 2002). The importance of PPR can be realized by the number of sheep and goats, which is now more than 1 billion in the countries where PPR has been reported at least once, compared to 750 million in 2002 (Diallo 2006). Bazarghani et al. (2006) estimated a loss of at least US$1.5 million to the Iranian farmers due to death of sheep and goats with PPR disease, with the cost of control measures far more than this (Bazarghani et al. 2006). A recent study conducted by (Thombare and Sinha 2009) estimated the series of factors that can influence the profit of farmers in the case of PPR outbreaks. Reduced market price of the diseased animal was the main factor, followed in descending order by loss in production, treatment costs, infertility, and labor services. The annual loss attributed to PPRV in Kenya is currently thought to be in excess of 1 billion Kenyan shillings (US$15 million).

Considering that there is one outbreak every 5 years in goats, it was estimated that the annual loss ranges from 0.57–3.92 dollars per animal in Nigeria, in which goats showed the most (3.92) and sheep the least (0.57) loss (Opasina and Putt 1985). Cumulatively, it has been estimated that PPRV causes a loss of 1.5 million US dollars annually, in Nigeria alone (Hamdy et al. 1976). The economic losses due to PPR in India have been estimated annually at 39 millions US$ (Bandyopadhyay 2002). In countries where PPR outbreaks occur, annual economic losses are in the range of millions of USD (Banyard et al. 2010). Despite low mortality and morbidity rates, recent outbreaks of PPRV in Morocco caused great economic concerns due to commercial trade between Morocco and both Algeria and Spain.

PPR disease is a major limiting factor not only for trade and export but also for the development of livestock production, especially small industries in developing nations. Financial crises might prevail in endemic areas, due to high mortalities, wiping out losses and the vaccination costs. PPRV can also deplete micronutrients and proteins from the affected animals, elements that are essential for human consumption (Turk 2009). Furthermore, PPRV can predispose animals to severe respiratory disease complex, especially in goats (Bailey et al. 2005; Taylor et al. 1990).

Peter Roeder, animal health officer responsible for viral diseases at FAO and the secretary of the global rinderpest eradication programme (GREP), states:

> It's becoming very clear that if you are undertaking any development of livestock production involving small ruminants in Asia and much of Africa, that PPR has to be taken extremely seriously. Livestock have to be protected against it.

## 7.3 Control and Eradication of PPRV: Where do We Stand?

There have been substantial improvements in designing efficient diagnostic assays and developing effective vaccines. Research in recent years has built solid foundations that can lead to global eradication of PPR, as practiced for RP. However, there are still several research gaps that need to be filled before promising any successful campaign for elimination of PPR.

### 7.3.1 Factors Favoring the Possibility for Global Eradication of PPRV

After the eradication of small pox from the world, an FAO and OIE joint report about the global eradication of RP appeared to be a milestone in the field of life sciences. Eradication of these two deadly diseases left hope and curiosity for the future of global eradication of epidemic diseases. PPR, in particular, is a suitable candidate because of identical etiological features, disease mechanism, and epidemiology to those of RP. Keeping these factors in mind, it is possible to repeat reliably the RP eradication strategies for PPR. It is plausible to conclude following 10 points, if not more, to be considered while evaluating the feasibility of global eradication of PPR.

### 7.3.1.1 Despite four Lineages, PPRV Strains Exist as a Single Serotype

Although PPRV can be classified into four lineages, based on either the F or N gene, fortunately there is only a single serotype of PPRV known so far. This means that a single vaccine prepared from any of the lineages will provide protection against all of the lineages. This leaves the possibility to apply a single vaccine to all of the affected animals without prior characterization of the viruses, and that this vaccine can conveniently be applied for mass immunization of the naïve host population. An attenuated tissue culture vaccine based on Nigeria/75/1 (Nig75/1), one of the very first isolates of PPRV, is widely used for vaccination and immunization of small ruminants in almost all of the PPRV endemic areas in the world. Additionally, vaccinated animals are unable to transmit the disease to nearby healthy flocks. This homologous vaccine appears to be safe for pregnant animals, and in field conditions induces protective immunity in at least 98 % of the vaccinated animals (Diallo et al. 1995).

This vaccine is currently being extensively used in the endemic areas of Africa, the Middle East, and Southeast Asia (Table 7.1). The Nig/75/1-based vaccine is available in CIRAD EMVT, at Montpellier, France, for all areas except Africa, for which the vaccine is available through PANVAC at Debre Zeit, Ethiopia. Besides Nig/75/1 and its other versions, three strains of PPRV named Sungri 96, Arasur/87, Coimbatore/97 are in use in India only (Sen et al. 2010) (Table 7.1). Another live attenuated vaccine from PPRV strain Egypt/87 is also available only in Egypt. Since only a single serotype exists for PPRV, all of these vaccines prove to be efficient in eliciting protective immunity against PPRV. It is, therefore, possible to arrange and apply coordinated and strategic mass vaccination for fruitful results, which is unfortunately currently lacking for most of the PPRV endemic countries.

### 7.3.1.2 Live Attenuated PPRV Vaccine Elicits Long Protective Immunity

The available live attenuated PPR vaccines, Nig/75/1 in particular, provide long protective immunity. It has been demonstrated that the Nig/75/1 vaccine protects the immunized small ruminants for a period of up to 3 years. Sungri 96, on the other hand, has been shown to sustain protection even up to 6 years. This long period of protection fully guarantees the efficient control and eradication of PPRV through mass vaccination, which immunizes the naïve goats or sheep.

Since PPRV is a disease of tropical countries, the thermostability remains a drawback of the live attenuated vaccines. Recently, it has been demonstrated that freeze-drying of this vaccine in an excipient-containing trehalose makes it very thermostable. This vaccine is resistant to temperature as high as 45 °C for 14 days, with negligible loss in efficacy (see Chap. 5, Sect. 3.3). The utilization of this vaccine to protect small ruminants would pave the way for a PPR control program, thereby making the option of vaccination under field conditions a technically viable and economically feasible solution, leading to effective control of PPR in developing countries.

**Table 7.1**  Availability of the PPRV vaccines and their application in PPR-endemic countries

| Countries using PPRV vaccines | Strain of PPRV used in the vaccine | Nature of the vaccine | Product name |
|---|---|---|---|
| Afghanistan, Albania, Bahrain, Ethiopia, Iraq, Jordan, Kuwait, Lebanon, Libya, Yemen, United Arab Emirates, Syria, Pakistan, and Oman | Nigeria 75/1 | Live modified | PESTEVAC |
| Botswana | PPRV 75/1 | Live | PPR-VAC |
| Egypt | Egypt 87 | Live | Not available |
| Nepal | PPRV 75/1 homologous | Live | Not available |
| Nigeria | PPRV 75/1 | Live | Not available |
| Turkey | PPRV Nigeria 75/1 | Live | PESTDOLL-S |
| India | PPRV Sungri 96 PPRV Arasur/87 PPRV Coimbatore/97 | Live | PPR-Vaccine |

### 7.3.1.3  PPRV Spreads Only Through Close Contact, Independent of Vectors

PPRV only spreads by close contact between sick animals and healthy animals. It does not rely on vectors such as *Culicoides*, which is required for bluetongue virus as the transmission medium. Therefore, control of PPR is comparatively easier than that of vector-borne diseases. In the case of PPRV outbreaks, it is possible to apply a stamping out strategy of the sick animals, which helps to reduce the spread of epidemics. It is difficult to wipe out the medium of transmission to an affected area, by killing the vector, thus making effective control problematic.

### 7.3.1.4  PPRV is Restricted to Small Ruminants and Camels

The host range for PPRV infection is restricted to small ruminants (sheep, goats, and wildlife). Recently, the involvement of camels among the susceptible host list made PPR a special disease. However, sheep and goats remain the main and primary host for PPRV infection. Compared to controlling a disease in chickens, ducks, and other poultry, small ruminants offer a larger and easier target to be focused upon. It is convenient, in the case of sheep and goats, to implement better husbandry and nutrition, to restrict the spread of an epidemic of PPRV. Moreover, being nonzoonotic, the process of dealing with sick animals is nonhazardous for human health, which leads to confident handling of the disease situation. However, due to the shorter working life of small ruminants, more vaccine and administration is required compared to cattle as was in the case of the RP eradication campaign.

### 7.3.1.5  Short Incubation Period of PPRV

The incubation period for PPRV ranges from 2 to 6 days, which depends upon the specific form of the disease. Additionally, there is no persistent and carrier state of PPRV reported. All of these elements favor the control of PPRV over many other viruses, which persist for a longer time and the infected animals remain in a carrier state.

### 7.3.1.6  PPR is not a Global Disease

Although, PPR is prevalent in most of the countries in Africa, the Middle East, and South Asia, the disease has not been reported in Europe, America, or Australia. However, there is an increasing trend in disease occurrence in previously PPRV-free countries. New outbreaks have been recorded in China and Morocco; however, both epidemics were relatively small, and the disease was effectively controlled in a short period of time. Given the relatively localized distribution of PPRV, it is imperative to restrict further spread of the disease.

### 7.3.1.7  Infected Animals Remain Seropositive

Infected and recovered animals remain seropositive for the specific antibodies against PPRV. The detection of these specific antibodies is the most common and economical way of detecting PPRV seropositive animals, which provides an efficient tool to seromonitor a non-vaccinated herd. However, unfortunately the currently available vaccines elicit the same kind of antibodies, which makes the differentiation of infected and vaccinated animals (DIVA) difficult, and therefore demands extensive research to develop recombinant vaccines with a DIVA strategy. We have recently reviewed all of the available serological techniques applied for the diagnosis of PPRV, and provided a comprehensive platform for improvements (Munir 2011; Munir et al. 2012).

### 7.3.1.8  Availability of Diagnostic Tests

Shifting attention to early warning, early detection, and early responses is the key to opening the door to efficient disease control. Diagnosis of PPR is usually made by clinical observation, and in typical cases the animals show characteristic signs and symptoms (see Chap. 3). However, due to the presence of aggravation factors or concurrent infections, the disease may be confused with several other diseases. In this case, serological or molecular confirmation is required.

Several ELISAs have been developed targeting either the HN (Anderson and McKay 1994; Saliki et al. 1993; Singh et al. 2004) or N proteins (Libeau et al. 1995) for specific detection of antibodies against PPRV, both in ovine and caprine hosts. Additionally, two different formats of ELISA have been developed and applied in the field to efficiently detect antigens in the tissues and secretions of PPRV infected animals. Imunocapture ELISA (Libeau et al. 1994) has overwhelmed sandwich ELISA (Saliki et al. 1994), but both have utilized MAbs directed against the N protein of PPRV. Moreover, different versions of reverse transcription (RT)-polymerase chain reaction (PCR) targeting the F protein or N protein mRNA have been developed (see Chap. 5, Table 5.1). Collectively, the traditional ELISAs and RT-PCR remain in-depth diagnostic techniques, which are now further advanced by the development of fluorescent quantitative RT-PCR, the LAMP, and immunochromatographic test strips.

### 7.3.1.9 Development of DIVA Vaccines

The traditional live attenuated PPRV vaccines are unable to distinguish between immunized and naturally infected animals (DIVA), because both basically induce the same type of antibody. A recombinant vaccine with DIVA capability accompanied by a suitable diagnostic test is required. Several advances have been made; however, a DIVA vaccine is still at the early stage of development and commercialization, and therefore requires immense research in the near future. Such a DIVA vaccine will be crucial to monitor infected and vaccinated animals at the onset of an eradication campaign for PPRV, and it is expected that this strategy will be available soon.

### 7.3.1.10 Development of Novel Recombinant and Multivalent Vaccines

PPRV, being significantly immunosuppressive, has been identified with other concurrent infections, notably bluetongue virus (BTV) (Mondal et al. 2009), sheep pox virus (SPV), goat pox virus (GPV) (Saravanan et al. 2007), and pestivirus (Kul et al. 2008). The geographical distribution of some of these diseases, such as GPV and PPRV, is similar, and developing countries demand economical infrastructures to support concerted vaccination programmes, so there have been significant advances in designing multivalent (bi- or tri-valent) vaccines for PPRV and other viruses that can ideally substitute for conventional live attenuated PPRV vaccines. Additionally, field application of these multivalent vaccines to control common pathogens would indeed help to enhance poverty alleviation. Considering the enormous benefits of multivalent vaccines, currently several vaccines are being developed that may protect vaccinated animals against several viral pathogens at the same time, such as PPRV, GPV, SPV, and BTV.

## 7.3.2 Factors Hindering the Possibility for Global Eradication of PPRV

Analysis of the factors favoring the global eradication of PPR has provided clues that global eradication is feasible and achievable. However, relative to the global rinderpest eradication programs, there are still several factors constraining the global eradication of PPRV. Analysis of these factors is mentioned below.

### 7.3.2.1 Economic Consequences of PPR are Underestimated

The significance of the PPR eradication from the globe cannot be completely realized without evoking its spearhead role in animal health. The economic losses associated with PPR are not only direct through reduction in animal production and high death rate but also indirect through trade losses due to restrictions on animal movements. Although it may be desirable to eradicate any health threat, cost-effectiveness is an important consideration, especially in developing countries, where public resources have many high priorities, and sustained expenditures require clearly discernable benefits for large segments of society. There have been several reports evaluating the direct losses; however, the indirect losses remained to be determined. Proper estimation of these costs is crucial to realize the impact of this devastating disease to initiate any control measures.

### 7.3.2.2 Unavailability of Vaccines in Some of PPR-Endemic Countries

Relative to the mass immunization for prevention and control of RP, the large-scale immunization for prevention of PPR is insufficient; even if there has been a lot of commercialization. This huge demand for vaccines for the small ruminant industry requires promoting the local production of live attenuated PPRV vaccines not only for immediate availability but also to reduce the cost of production. Moreover, provided that PPR is a disease of tropical countries, development of thermostable vaccines with improved shelf life warrants further investigations.

### 7.3.2.3 Roles of Wildlife and Camels in the Epizootiology of PPR Remain Elusive

Although, there have been several reports on the pathogenicity of PPRV in the primary hosts, sheep, and goats, the role of wildlife and camels in the epizootiology is not completely understood. Since some wild animals play crucial roles in the spread of PPRV and may lead to epidemics, it is highly desirable to study the viral biology in these animals. Such findings are also important in devising control strategies and planning eradication of PPRV.

### 7.3.2.4  Ineffecient Serosurveillance System for PPR

An efficient post-outbreak serological monitoring system is an important tool in the eradication of a disease. However, as described above, both immune protection and serological monitoring in parallel require a DIVA vaccine and a suitable companion diagnostic test, which are currently lacking but are at the initial stage of development. Additionally, strengthening the infrastructures for diagnostics and epidemiology is the key issue in many developing countries. The slow development of this approach needs to be facilitated so that the true epidemiological investigation and planning for the restriction of the disease can be made as early as possible.

### 7.3.2.5  Definition of Disease Free Zones is Lacking in Many Countries

The management of a specific animal disease requires defining the disease-free zone within a country, as recommended by the OIE. Although this requires wider application, China implemented and specified animal disease-free zone standards in 2008, which has helped China to not only monitor the spread of the disease but also to provide a foundation to mark the disease-prone areas to be monitored for transportation and quarantine. However, due to technical, economical, and political issues, it is difficult for some developing countries to practice this approach, which requires promotion by international organizations.

### 7.3.2.6  Globalization and Animal Diseases

The concept of a "global village" and a range of other factors are contributing significantly to the spread of diseases, and it is becoming harder and harder for countries to maintain a disease-free status. These factors include: (i) demographic factors, (ii) global public mobility, (iii) significant increases in mobility and transport of live animals across the globe, (iv) trade and trafficking of animals, especially between neighboring countries at semi-controlled borders, (v) transportation of animal products, (vi) climatic changes and their overall impact on disease patterns, (vii) an increase in the middle-class livestock sector, (viii) an increase in human-animal contact, (ix) dynamics of food and agriculture, and (x) unsustainable resource management. Although the layout for efficient management of these aggravating factors is not currently defined, it remains a dilemma that these factors are a great hindrance to the prevention and elimination of any emerging disease such as PPR.

## 7.4 Roles of International Health Organizations in PPR Eradication Campaigns

A unified framework is currently lacking that can help bridge and synthesize the lessons from the RP eradication programme, and effectively devise future campaigns for the successful control and elimination of PPR from the globe. International organizations, including FAO and OIE, should play both facilitatory and leading roles in (i) promoting and securing solid efforts toward global eradication of PPR, (ii) assisting and harmonizing the local, regional, and global inputs, (iii) permitting and streamlining basic veterinary services across different ecosystems, (iv) bringing countries together on a common platform by overcoming the boundaries and hurdles in animal health technologies, and (v) providing financial and technical assistance. In the GREP, originated by FAO and OIE, a major chunk of the funding was utilized for the development of infrastructures. It is therefore pertinent to consider the possibility of eradicating PPR in the same frame. Considering the economic impact of PPR in the small ruminant industry and the preparedness in terms of vaccines and diagnostic tests, it is worthwhile and feasible for the FAO and OIE to consider an eradication programme for PPR in the near future.

There have been some initiatives by FAO in which strategies to eradicate PPR have been proposed, which consist of awareness campaigns for the governments and stakeholders, establishing global and regional policies, efforts to understand the ecology and epidemiology of PPR, the use of efficient and thorough vaccination, and collaboration with other ongoing campaigns. Moreover, in the context of socioeconomic effects, international support, and funding will absolutely be required to eradicate the PPR from the globe.

Besides FAO and OIE, it is expected that the institute of animal health (IAH), UK, will play a decisive role in a PPR eradication campaign through on-going research programmes, such as researchers at the IAH under the leadership of Dr. Barret (now deceased) have played in the RP eradication. Collaboration of IAH with the Centre de Coopération Internationale en Recherche Agronomique poure le Développement (CIRAD), France, and other non-European laboratories would bring an effective, practical and need-based research programme that would eventually build a solid platform to eradicate PPR. In particular, the importance for the involvement of researchers from PPR-endemic countries in any strategy and planning for PPR should not be overlooked. Recently, there has been an unofficial meeting at the IAH, under the leadership of Dr. Baron, for exchange of current advances in PPR research, in which representatives of limited countries participated (Dr. Valarcher, National Veterinary Institute, Sweden, personal communication). We consider this as an important initiative, and are expecting to have a bigger gathering in the near future, with participation of people from the unrepresented countries. A systematic programme to address global PPRV eradication is currently missing, and urgently requires the support from government, international organizations and funding agencies.

## 7.5 Conclusion and Prospects

Vaccination, combined with common sense actions such as safeguarding flocks and herds against inadvertent introduction of the disease, is able to provide good control, particularly as a vaccine is available to provide life-long immunity. But, if this can be achieved within a programme of progressive control, losses could be minimized and certain areas should be freed from PPR. In addition to the strict management of the infected area of animals, non-affected areas should be marked and monitored frequently, small ruminants in the border areas should be sero-monitored, and there is a need to strictly apply control measures for the import and export of animal products and trade. Despite the fact that a single serotype exists, it is imperative to use of prototype in each lineage for vaccine production, which should be used in the areas where the specific lineage is prevalent. The primary reason for this region-specific vaccine is the tendency of PPRV, being an RNA virus, to mutate extensively. Besides this, the combined vaccine produced at a large scale and applied in the field will be a very cost-effective alternative to individual vaccination strategies in developing countries.

The existing technical tools and animal health systems provide a solid foundation for initiating progressive control operations of this disease on small ruminants. Unless coordinated actions are taken by international organizations to control the spread of the disease, PPR is likely to spread to the rest of the African and Asian countries, bringing with it untold losses of livestock and endangering the livelihoods of millions of farmers and herders. Before eradication is initiated, further support should be provided by FAO to help authorities understand PPR, and to be able to differentiate it from a variety of diseases that cause similar respiratory problems and mortality of small ruminants, including pneumonic pasteurellosis and contagious caprine pleuropneumonia. Collectively, the evaluation of economic impact, improvement, and commercialization of diagnostic tests and vaccines, and coordination and integration for planning are key elements to be considered in PPR eradication.

## References

Anderson J, McKay JA (1994) The detection of antibodies against peste des petits ruminants virus in cattle, sheep and goats and the possible implications to rinderpest control programmes. Epidemiol Infect 112(1):225–231

Arzt J, White WR, Thomsen BV, Brown CC (2010) Agricultural diseases on the move early in the third millennium. Vet Pathol 47(1):15–27

Awa DN, Njoya A, Ngo Tama AC (2000) Economics of prophylaxis against peste des petits ruminants and gastrointestinal helminthosis in small ruminants in North Cameroon. Trop Anim Health Prod 32(6):391–403

Bailey D, Banyard A, Dash P, Ozkul A, Barrett T (2005) Full genome sequence of peste des petits ruminants virus, a member of the Morbillivirus genus. Virus Res 110(1–2):119–124

Bandyopadhyay SK (2002) The economic appraisal of PPR control in India. In: 14th annual conference and national seminar on management of viral diseases with emphasis on global trade and WTO regime, Indian Virological Society, Hebbal, Bangalore, India, 18–20 Jan 2002

Banyard AC, Parida S, Batten C, Oura C, Kwiatek O, Libeau G (2010) Global distribution of peste des petits ruminants virus and prospects for improved diagnosis and control. J Gen Virol 91(Pt 12):2885–2897

Bazarghani TT, Charkhkar S, Doroudi J, Bani Hassan E (2006) A review on peste des petits ruminants (PPR) with special reference to PPR in Iran. J Vet Med 53(Suppl 1):17–18

Diallo A (2006) Control of peste des petits ruminants and poverty alleviation? J Vet Med 53(Suppl 1):11–13

Diallo A, Libeau G, Couacy-Hymann E, Barbron M (1995) Recent developments in the diagnosis of rinderpest and peste des petits ruminants. Vet Microbiol 44(2–4):307–317

Ezeokoli CD, Umoh JU, Chineme CN, Isitor GN, Gyang EO (1986) Clinical and epidemiological features of peste des petits ruminants in Sokoto Red goats. Revue d'elevage et de medecine veterinaire des pays tropicaux 39(3–4):269–273

FAO (2008) Outbreak of 'peste des petits ruminants' in Morocco. FAO Newsroom (FAO), 9 September

Hamdy FM, Dardiri AH, Nduaka O, Breese SRJ, Ihemelandu EC (1976) Etiology of the stomatitis pneumcenteritis complex in Nigerian dwarf goats. Can J Comp Med 40:276–284

Kul O, Kabakci N, Ozkul A, Kalender H, Atmaca HT (2008) Concurrent peste des petits ruminants virus and pestivirus infection in stillborn twin lambs. Vet Pathol 45(2):191–196

Kwiatek O, Minet C, Grillet C, Hurard C, Carlsson E, Karimov B, Albina E, Diallo A, Libeau G (2007) Peste des petits ruminants (PPR) outbreak in Tajikistan. J Comp Pathol 136(2–3):111–119

Libeau G, Diallo A, Colas F, Guerre L (1994) Rapid differential diagnosis of rinderpest and peste des petits ruminants using an immunocapture ELISA. Vet Rec 134(12):300–304

Libeau G, Prehaud C, Lancelot R, Colas F, Guerre L, Bishop DH, Diallo A (1995) Development of a competitive ELISA for detecting antibodies to the peste des petits ruminants virus using a recombinant nucleoprotein. Res Vet Sci 58(1):50–55

Mondal B, Sen A, Chand K, Biswas SK, De A, Rajak KK, Chakravarti S (2009) Evidence of mixed infection of peste des petits ruminants virus and bluetongue virus in a flock of goats as confirmed by detection of antigen, antibody and nucleic acid of both the viruses. Trop Anim Health Prod 41(8):1661–1667

Munir M (2011) Diagnosis of peste des petits ruminants under limited resource setting: a cost effective strategy for developing countries where PPRV is endemic. VDM Verlag Dr. Müller, Germany

Munir M, Abubakar M, Zohari S, Berg M (2012) Serodiagnosis of peste des petits ruminants virus. In: Al-Moslih M (ed) Serological diagnosis of certain human, animal and plant diseases, vol 1. InTech, pp 37–58

Nanda YP, Chatterjee A, Purohit AK, Diallo A, Innui K, Sharma RN, Libeau G, Thevasagayam JA, Bruning A, Kitching RP, Anderson J, Barrett T, Taylor WP (1996) The isolation of peste des petits ruminants virus from northern India. Vet Microbiol 51(3–4):207–216

Opasina BA, Putt SNH (1985) Outbreaks of peste des petits ruminants in village goat flocks in Nigeria. Trop Anim Health Prod 17:219–224

Perry BD, Randolph TF, McDermott JJ, Sones KR, Thornton PK (2002) Investing in animal health research to alleviate poverty. International Livestock Research Institute, Nairobi

Rossiter PB, Taylor WP (1994) Peste des petits ruminants. In infectious diseases of livestock, vol II. Oxford University Press, Cape Town

Saliki JT, House JA, Mebus CA, Dubovi EJ (1994) Comparison of monoclonal antibody-based sandwich enzyme-linked immunosorbent assay and virus isolation for detection of peste des petits ruminants virus in goat tissues and secretions. J Clin Microbiol 32(5):1349–1353

Saliki JT, Libeau G, House JA, Mebus CA, Dubovi EJ (1993) Monoclonal antibody-based blocking enzyme-linked immunosorbent assay for specific detection and titration of peste-des-petits-ruminants virus antibody in caprine and ovine sera. J Clin Microbiol 31(5):1075–1082

Saravanan P, Balamurugan V, Sen A, Sarkar J, Sahay B, Rajak KK, Hosamani M, Yadav MP, Singh RK (2007) Mixed infection of peste des petits ruminants and orf on a goat farm in Shahjahanpur, India. Vet Rec 160(12):410–412

Sen A, Saravanan P, Balamurugan V, Rajak KK, Sudhakar SB, Bhanuprakash V, Parida S, Singh RK (2010) Vaccines against peste des petits ruminants virus. Expert Rev Vaccines 9(7):785–796

Singh RP, Sreenivasa BP, Dhar P, Shah LC, Bandyopadhyay SK (2004) Development of a monoclonal antibody based competitive-ELISA for detection and titration of antibodies to peste des petits ruminants (PPR) virus. Vet Microbiol 98(1):3–15

Stem C (1993) An economic analysis of the prevention of peste des petits ruminants in Nigerien goats. Prev Vet Med 16:141–150

Taylor WP (1984) The distribution and epidemiology of PPR. Prev Vet Med 2:157–166

Taylor WP, al Busaidy S, Barrett T (1990) The epidemiology of peste des petits ruminants in the Sultanate of Oman. Vet Microbiol 22(4):341–352

Thombare NN, Sinha MK (2009) Economic implications of peste des petits ruminants (PPR) disease in sheep and goats: a sample analysis of district Pune, Maharastra. Agric Econ Res Rev 22:319–322

Turk J (2009) Global food crisis: the value of animal source foods. ELMT/ELSE Newsletter

Wang Z, Bao J, Wu X, Liu Y, Li L, Liu C, Suo L, Xie Z, Zhao W, Zhang W, Yang N, Li J, Wang S, Wang J (2009) Peste des petits ruminants virus in Tibet, China. Emerg Infect Dis 15(2):299–301

# Index

M. Munir et al., *Molecular Biology and Pathogenesis of Peste des Petits Ruminants Virus*, SpringerBriefs in Animal Sciences, DOI: 10.1007/978-3-642-31451-3, © The Author(s) 2013